The Politics of Procurement

Aaron Plamondon

The Politics of Procurement
Military Acquisition in Canada and the *Sea King* Helicopter

UBCPress · Vancouver · Toronto

21 20 19 18 17 16 15 14 13 12 11 10 5 4 3 2 1

Printed in Canada on FSC-certified ancient-forest-free paper (100 percent post-consumer recycled) that is processed chlorine- and acid-free.

Library and Archives Canada Cataloguing in Publication

Plamondon, Aaron, 1975-
The politics of procurement : military acquisition in Canada and the Sea King helicopter / Aaron Plamondon.

Includes bibliographical references and index.
ISBN 978-0-7748-1714-1

1. Canada. Canadian Armed Forces – Procurement. 2. Canada – Armed Forces – Procurement. 3. Defense contracts – Government policy – Canada. 4. Canada. Canadian Armed Forces – Equipment. 5. Canada – Armed Forces – Equipment. 6. Military helicopters – Purchasing – Canada. 7. Sea King (Helicopter). 8. Defense contracts – Canada. I. Title.

UA600.P53 2010 355.6'2120971 C2009-905204-0

Canadä

UBC Press gratefully acknowledges the financial support for our publishing program of the Government of Canada through the Book Publishing Industry Development Program (BPIDP), and of the Canada Council for the Arts, and the British Columbia Arts Council.

This book has been published with the help of a grant from the Canadian Federation for the Humanities and Social Sciences, through the Aid to Scholarly Publications Programme, using funds provided by the Social Sciences and Humanities Research Council of Canada.

Cover image: *Top*: A sunrise view of a CH-124 *Sea King* on the tarmac at 438 Tactical Helicopter Squadron (438 Tac. Hel. Sqn.). *Bottom*: Operation Sextant, Indian Ocean: Captain Joel MacDermaid, a Navigator from 443 Maritime Helicopter Squadron. *Sources*: CF photos by Sgt. René Dubreuil and by Warrant Officer Carole Morissette. *Courtesy of National Defence. Reproduced with the permission of the Minister of Public Works and Government Services, 2009*

UBC Press
The University of British Columbia
2029 West Mall
Vancouver, BC V6T 1Z2
604-822-5959 / Fax: 604-822-6083
www.ubcpress.ca

Contents

List of Illustrations

Operation ALTAIR: A CH-124 *Sea King* helicopter conducts a wet hoist exercise on 28 May 2008.

Source: CF photo by M.Cpl. Robin Mugridge. *Courtesy of National Defence. Reproduced with the permission of the Minister of Public Works and Government Services, 2009*

Preface

> When the Avro *Arrow* was cancelled, at least there was a study involved in the decision. The cancellation of the *Sea King* replacement involved no study at all, and this was really an unforgivable error. It was our worst national procurement mistake ... worse than the *Arrow*.
>
> – Former Chief of the Defence Staff General (retired) Paul Manson

The procurement of military weapons and equipment in Canada has often been controlled by partisan political considerations – not by a clear desire to increase the capability of the military. Actual military strength has typically been given a low priority by Canada's civilian leaders. As a result, Canada has often failed throughout its history to be effective in the design, production, or even purchase of weapons and equipment necessary for its military to carry out the priorities of the civil power. To secure even the most modest materiel, military officials have had to comply with a succession of rules that can only be described as illogical from the standpoint of military performance. Much like Alice in *Through the Looking Glass,* they have had to run twice as fast just to stay in the same place. National Defence Headquarters (NDHQ) has not helped itself, however, by making the process more efficient from its end. The internal process has continually evolved into an amorphous mass of bureaucracy, with a myriad of committees that require endless analyses, re-evaluations, and approvals. The history of the *Sea King* maritime helicopter and the failed attempts to replace it over three decades is a product of all these weaknesses within the Canadian procurement system. It is the ultimate case study.

This is a story of delays. The word, and all its possible synonyms, will, in fact, be overused. There is simply no way around it. This study reveals the increasing timeline of procurement in Canada by tracing the introduction of naval helicopters in the 1950s, the procurement of the first modern Anti-Submarine Warfare (ASW) helicopter, the Sikorsky *Sea King* in 1962, and the multiple failed attempts to replace it. As I will explain in the Introduction, the procurement process evolved over time and used varying nomenclature but was usually composed of several fundamental elements: a definition of military requirement; validation of the requirement; government approval of the project; selection of a procurement strategy; bid solicitation and source selection; negotiation and award of contract; administration of the contract to purchase the piece of equipment decided on; delivery of the product, life cycle support, and eventually disposal.[1] Although these phases have been present throughout the period of this study, the time that it took to complete them continually increased. Early helicopter acquisition needed only a few pages to state the requirement; by the time that the New Shipborne Aircraft (NSA) project was initialized in 1985, these documents were referred to in volumes. Discussions on the replacement of the *Sea King* helicopter fleet officially began in 1975. By the time that this book is completed, it will still be years away. It has now been a decade since military analyst Joseph Jockel surmised that "the Sea Kings are now operationally useless, except for helicopter training and providing some utility lift, largely because their sensors are outdated and very difficult to keep functioning at all."[2] The fact that the attempts to replace the *Sea King* necessitate a book-size study reveals that there is a procurement problem in Canada. The fact that the story is ongoing reveals that the problem is severe. It could be the most poorly executed military procurement ever undertaken – anywhere.

Helicopters have a long and proud history in Canadian naval operations. The use of a medium-sized helicopter aboard smaller Destroyer Escort (DE) vessels instead of aircraft carriers in the 1950s was a distinctly Canadian idea. The technical innovations used by Canadian engineers to carry this out changed how ASW operations were conducted by every modern navy in the world, including those of Britain and the United States. There are few examples, indeed, where Canada can lay claim to such initiative and ingenuity regarding military matters. As Canada had only one carrier, naval officers believed that, if they could fly helicopters from their smaller vessels, they could maximize the potential of their ships. These vessels were already lagging behind Soviet submarine technology. Discussions on the procurement of a fleet of ASW helicopters for the Royal Canadian Navy (RCN) had been ongoing since 1954, and by 1961 officials were still wavering over which aircraft to buy. This waffling was due largely to the uncertainty of the role that helicopters were going to fulfill in the RCN and rivalries between the three Canadian services, army, air force, and navy. A further complication with the introduction of helicopters into the RCN was the completion of trials for the Hauldown and Rapid Securing Device, or *Beartrap,* as it was more commonly known. The *Beartrap* was the integral component of the helicopter-carrying DE platform that allowed a large helicopter to land on such a small ship in the turbulent conditions common in naval operations. It was designed and built in Canada. As it was a new technology, it was necessary to put it through extensive testing before helicopters could safely fly from destroyers. The concept had been conceived in 1955, and the design, production, and development trials were carried out throughout the 1960s. The *Beartrap* was later patented in Canada.

The *Beartrap* formed the basis for the procurement of the first advanced ASW helicopter in Canada – the Sikorsky *Sea King.* This aircraft was acquired to counter the Soviet submarine threat during the Cold War. After its introduction into the RCN, it quickly mitigated the problem of a lack of modern Canadian ships to carry out the ASW commitments within the North Atlantic Treaty Organization (NATO) that Canada had accepted in 1949. As Martin Shadwick has written, "a formidable ASW tool in its own right, the Sea King breathed new life into surface vessels which were clearly outclassed by the latest nuclear-powered submarines."[3] Marc Milner agreed: "The Sea King was a big, powerful machine ideally suited to the ambitious shipboard helicopter concept developed by the RCN."[4] The first *Sea King* landed in Canada in 1963, and the last was received in 1969. In the same year that the last *Sea King* entered service in Canada, it was announced that the RCN's only aircraft carrier, Her Majesty's Canadian Ship (HMCS) *Bonaventure,* was to be retired from service. This would have a profound effect on ASW helicopter operations. As fixed-wing aircraft were phased out at sea along with the *Bonaventure,* the *Sea King* fleet was to make up for their absence.

The ASW capability of the *Sea King* began to lag behind subsurface technology in the 1980s. The *Sea King* was to be replaced in the early 1990s, and it was already clear that naval helicopters would be used for much more than ASW in the post-Cold War era. They were to contribute to the maintenance of Canada's sovereignty and security through active participation in drug interdiction, environmental monitoring, fisheries protection, search and rescue, international peacekeeping and humanitarian aid, control of maritime approaches, and response to international contingency operations, including support to land force operations. The early 1990s saw the *Sea King* deployed in new support roles in the Gulf War, Somalia, Haiti, and the former Yugoslavia. These new roles did not mean, however, that the submarine threat had vanished from international waters or that Canada did not need an ASW capability aboard its naval helicopters. When the Statement of Requirement (SOR) was being written in the 1980s, ASW was still the primary focus. But the strategy of the Department of National Defence (DND) in the early 1990s was to re-equip the Canadian Forces (CF) with adaptable and effective equipment that could perform a variety of tasks over the course of its service life. Flexibility was essential. It was not certain that the increasing number of countries with sophisticated naval capabilities would behave in a way that was acceptable to Canada or its allies. As a result, Canada's maritime forces needed to be able to operate effectively in a modern naval environment. Defence planning has always been about being prepared to counter future threats, and the replacement of the maritime helicopter capability was overdue by the 1990s. The *Sea King* was no longer effective in the new security environment.

In 1987, the *EH-101* helicopter built by European Helicopter Industries (EHI) was announced the winner of the NSA competition to operate aboard the Canadian patrol vessels then under construction. EHI was a British and Italian consortium established with the sole purpose of creating a replacement for the *Sea Kings* in those navies. The *EH-101* was also selected by the Canadian government because it was the best helicopter available at the time and would complement the new ships in the maritime role. Much like the original *Sea King,* the *EH-101* was selected on merit. Although both acquisitions took far longer than expected to enter into the contract phase, it was obvious at the respective times of selection that both helicopters would increase the capabilities of the navy exponentially. But the selection of the *EH-101* model by the government in 1987 was not the final phase of the NSA procurement process; in fact, it was really just the beginning. Although it took over a decade to choose a helicopter after the creation of the original SOR by the military, this event marked only the beginning of a new process of work and investment that was never to have a result.

EHI had presented the best bid with the best product. The 1987 Defence White Paper, however, had determined that the protocol for major Canadian defence acquisitions necessitated the inclusion of Canadian industry in their completion. The policy of securing Industrial and Regional Benefits (IRBs) as part of major defence contracts had become most explicit with the procurement of the long-range patrol aircraft (LRPA) in 1976. The procurement of the CP-140 *Aurora* went beyond the traditional "Canadian content" provisions and included specific contractual obligations on the company, Lockheed Aircraft Corporation, to attempt to achieve a wide variety of economic objectives in Canada over the lifetime of the program.[5] The company had to "offset" the cost to Canada by somehow contributing to the national economy and facilitating industrial involvement in the project. Offset policies were certainly not unique to Canada.[6] The procurement of the *Leopard I* tank and the *CF-18* jet fighter later followed this theme. The contract for the Canadian Patrol Frigate (CPF), awarded in June 1983, also aspired to maximize the participation of Canadian industry. It went one step further, however, and dictated to industrial bidders that any foreign contractor would have to form a consortium with a Canadian company, which would then lead the project. The goal, of course, was to stimulate the Canadian defence industry. The 1987 Defence White Paper continued the policy of tying domestic economic development to military procurement.

Although defence policy must clearly be based on political considerations, defence purchasing should be based primarily on military capability to allow the forces to carry out their mandate. The Canadian procurement policy that began in 1976 and continued in 1987 prevented expedient acquisitions by focusing on non-military considerations. The result was that, after the NSA bid was accepted, it took another five years to determine how Canadian industry would be involved in the process. This was called the "Contract Definition Phase," and EHI did not make its final bid to the government on how it would be implemented until 1992.

EHI's first offer to the government on how it would include Canadian industry was rejected in 1990 because it was considered too expensive. At approximately the same time, the government was considering whether to combine the NSA project with the New Search and Rescue Helicopter (NSH) project. This merger subsequently became a reality. The decision was based on the economic advantages that could be achieved if only one aircraft was used for both roles. This was commonly called "fleet rationalization." It was to involve a joint life cycle cost savings and a joint purchase through a common program office. But the merger of the two projects also meant that EHI had to re-establish its implementation plan based on new numbers and requirements. Government policy demanded that the solutions to these new requirements be resolved by Canadian industry. This policy obviated the purchase of the models already established by the British and the Italians. The NSA, therefore, returned into limbo, and the focus on the IRB policy hindered further progress on the NSA/NSH project. It was determined by EHI and the government that the company would provide the airframe but that the electronic mission suite inside would be assembled by Canadian industry. Some of the electronics had not yet been designed. It was speculated that approximately 400 companies were to be involved in the NSA project.

In July 1992, the government announced the award of a contract to EHI for the provision of fifty *EH-101*s – thirty-five for the NSA and fifteen for the NSH requirement. The NSA version was to be called the CH-148 *Petrel* and the NSH the CH-149 *Chimo*. The total cost was $4.4 billion. Although the project had been part of the public record since its beginnings in 1985, it was not until the summer of 1992 that there was any serious challenge to it by the Liberal opposition. By early 1993, the Conservative party in power had begun to fracture, and the economy was in a recession. At the best of times, defence expenditures in Canada are largely unpopular. In a time of fiscal restraint, a multi-billion-dollar deal to purchase military equipment was an easy target for the Liberals. Savvy politicians, led by Jean Chrétien, deemed the contract a waste of money.

The project became a topic primarily because it had taken too long to sign the final contract for delivery. This delay was due to the government policy of designing, producing, and assembling a Canadian mission suite into a foreign aircraft. As Canada entered into an election year, one political decision had created the opportunity for another. The project had gone on for too long with too few results. The political opposition seized on the inability of the party in power to convince the electorate of its necessity. Attempts to replace the *Sea King* had been under way since the 1970s, and the *EH-101* had been unanimously determined by the experts after extensive analysis to be the best model. None of its competitors, whether certified or in the design phase, came close to matching its value for money. But this was either never understood or ignored by Canadian citizens. It is far more likely that most Canadians did not comprehend the true nature of the purchase as a result of the effective Liberal campaign of misinformation. The opposition employed a dual strategy of exaggeration. First, they attacked the cost of the program, extrapolated the numbers to fit their objective, and refused to acknowledge that the final cost of the program was not just for the helicopter but also for its lifetime costs of maintenance. They also insinuated that the entire amount was to be paid immediately by the DND instead of the actual thirteen-year budget disbursement. Second, they created the illusion that the *EH-101* was an "attack" helicopter used solely to hunt Russian submarines and persisted in the assertion that, in the peaceful times of the post-Cold War era, this capability was not needed. The argument was simple: after the NATO treaty was signed and the Canadian government agreed to a central role in ASW, the acquisition of a fleet of modern naval helicopters designed for that purpose was justified. With the end of the Cold War, the Liberals extolled to the public how peace and stability would be the norm and removed the primary rhetoric for shipborne ASW and the replacement of the *Sea King*.

On 25 October 1993, the people of Canada voted for a government that they knew would cancel the helicopter replacement. There were, indeed, other issues at hand. Rarely do defence issues shift the momentum of an election in Canada. Only in 1911 and 1963 was defence an issue of magnitude. In the 1911 election, Prime Minister Sir Wilfrid Laurier was forced to defend his earlier establishment of a distinct RCN. The 1963 election centred on Prime Minister John Diefenbaker's refusal to fulfill commitments that his government had made through NATO and the more recently signed North American Air Defence Command (NORAD) to acquire nuclear warheads for weapons systems that Canada had already bought.[7] And in 1993, cancelling a major defence contract was part of the official Liberal platform; it was something that Chrétien said he would do, and when he was elected he did it. He cancelled the NSA/NSH within hours of officially taking power. It was an ostentatious demonstration to show that he would be a man of his word – that he would not hesitate to carry out the promises that he had made. Chrétien gave the impression that he would lead through action. The total costs of the contract termination amounted to $478.3 million.[8] The loss of work and investment to the Canadian defence industry, including the millions invested into research and development, was far more economically damaging. Workers across the country were laid off the day after the cancellation.

David Bercuson has written that, "of all the interesting, dramatic, exciting[,] aspects of defence policy and military operations, none is more dull than procurement. The very word seems to induce boredom."[9] Not so with the *Sea King* saga. The CF's attempt to replace the *Sea King* has ultimately turned into the worst procurement failure in Canadian history, even surpassing the ignominy of the A.V. Roe (Avro) *Arrow* cancellation of 1959. The intrigue of this ongoing story will only continue as more information that is now hiding or classified becomes available. As former *Sea King* pilot Colonel (retired) John Cody has expressed, "this is a subject that is so vast that it will take historians fifty years to sort it all out."[10] Indeed, there has even been an attempt by the Liberal government to remove any possibility of historical criticism by destroying all documentation that related to the NSA. Colonel (retired) Laurence McWha was in charge of collecting all of the contract documents and paper deliverables at the base in Shearwater at the time and recounted that "I called the Project Office in Ottawa for shipping info. A directive was eventually received to destroy all NSA documents. They went to the Base incinerator."[11]

The research for this study has not been easy. Nor will this book be the final word on the NSA cancellation. But it will demonstrate many important themes. Central to this demonstration is that weapons and equipment procurement in Canada has historically been an inefficient process, and the acquisition of a capability for the military has often been secondary to political considerations. When these political factors are the focus, they drastically extend the timeline of acquisition. And in Canada, the longer a procurement takes, the more politically vulnerable it becomes. The cancellation of the *Arrow* project and, later, that of the NSA, have made this clear. Former Chief of Defence Staff General (retired) Paul Manson has put it succinctly: "The Maritime Helicopter projects appear to reinforce the common perception that there has been inordinate growth in procurement times for military equipment in Canada over the past several decades."[12] After all the years of analysis on what was needed in a new naval helicopter, which model would be best suited, and how Canadian industry would be involved, the analysis turned to how much it would cost to scrap the whole thing.

After over a decade of preparation, the attempts to replace the *Sea King* were thwarted by political opportunism. The cancellation had little to do with discussions of compatibility with or capability for the navy. Indeed, there was no discussion between the Liberal government and its military advisers on the NSA project. As Colonel Cody has asserted, "I think history will tell the rest of the story, which was that the Government didn't give a hoot about us in particular or the Canadian Military in general."[13] Chrétien had, in fact, made it clear during the 1993 cancellation that he would not "lose any sleep over it."[14]

The following year the Defence White Paper reiterated the fact that there was an urgent need to replace the *Sea King,* and the DND would begin immediate work on the procurement of new helicopters. They were to enter Canadian service before 2000. It also stressed the purchase of off-the-shelf equipment to avoid technologies still under development. The first step, as always, was for the military to define what it needed in an SOR. It was to be based on the White Paper, which stated that the military needed to be a multi-role, combat-capable force able to fulfill a series of other DND mandates, such as assistance in protection of the fisheries, drug interdiction, environmental protection, humanitarian and disaster relief, and potential demands for aid to the civil power.

Although any replacement of the maritime helicopter role would obviously need robust capabilities to carry out this variety of tasks, some of the senior officers in charge of the procurement were afraid to base the SOR strictly on military requirements. They had been warned by their political masters not to exceed minimum requirements. These officers hardly needed a warning, however, as they understood how the cancellation of the NSA/NSH had politicized the issue. They knew that, if they created an SOR that led to the most capable aircraft, still thought to be the *EH-101,* the government would balk at the acquisition to avoid the embarrassment of buying the same helicopter that they had already renounced as unnecessary. The result was that the SOR had to be constantly rewritten to make it politically acceptable. The necessary capabilities of the maritime aircraft were reduced by 25 percent. The idea of creating an SOR not based on complete threat assessments and what the helicopter would be required to do, and where it would have to do it, was opposed by many at the DND.

It was decided by the government that the Search and Rescue (SAR) helicopters would be replaced first, and the SOR for the new Canadian Search and Rescue Helicopter (CSH) was completed in the summer of 1995. The *EH-101* won again. It was then being called the *Cormorant.* Although the government tried to avoid making the decision through various independent assessments regarding the validity of the competition, it was forced to concede that the *EH-101* was still the best aircraft or face a powerful legal battle with EHI. The political fallout over the CSH procurement served as an example of what had to be avoided for the *Sea King* replacement, by then being called the Maritime Helicopter Project (MHP). The government would require as many helicopters as possible to compete, and the only way to do that was to intervene in the acquisition process before the release of the Request for Proposal (RFP) to industry. It continued to reduce the Requirement Specifications (RS) for the aircraft that determined the final form of the RFP to allow less capable helicopters to compete. The final SOR was not complete until the summer of 1999 – five years after the first draft. The weaknesses of this portion of the procurement are shared by the DND as it was up to that department to submit the documents to the government for review. Although its trepidation over the SOR was justified, it only contributed to the already protracted procurement timeline.

The political pressure on the military and MHP office to reduce the RS was only one tactic available to the government. It also had the authority to decide the form that the final procurement strategy would take. The Liberals then created a committee chaired by Deputy Prime Minister Herb Gray to assess the process. As Colonel Brian Akitt, a former *Sea King* pilot and project director of the MHP, explained, "the introduction of the Gray Committee ensured that the Government had intervention into the process at the Departmental level thereby ensuring that the choice of procurement strategy and the definition of requirements would no longer fall within the purview of the Department. The military component of the relation was effectively neutralized."[15]

On 17 August 2000, the Chrétien government gave the DND official approval to proceed with acquiring a replacement for the *Sea King*. It had been decided to split the contract between a company that would build the airframe and another that would build and integrate the internal mission systems. It was explained that this unconventional strategy was chosen to broaden the list of companies able to compete for the project. The hope was that, with more companies involved, it would be less likely that the competition would once again lead to a variant of the *EH-101*. But it also meant that the project office had to establish two sets of requirements in anticipation of dealing with two separate prime contractors. This political decision increased the risk of massive delays and cost overruns due to the modifications that would be necessary to incorporate the electronics inside the airframe and then recertify the aircraft. This procurement strategy was questioned by senior officials in both the DND and Public Works and Government Services Canada (PWGSC), which was responsible for tendering all major national contracts. It also ignored the lesson of the NSA/NSH of incorporating too many companies into one acquisition.

Perhaps the most important decision regarding the procurement process that would mitigate the strength of the then named AgustaWestland International Limited (AWIL) *Cormorant* bid was the government's choice to use a "lowest-cost-compliant" methodology. This decision meant that, as long as a company was compliant with the RFP, if it submitted the lowest bid, then it would be declared the winner. This matrix contradicted both the PWGSC and the Treasury Board Contracting Policy of purchasing the best value. It also directly affected the competition as the *Cormorant* was a more capable and certified aircraft with more future potential than the rest of the field – but it was also more expensive. The split procurement based on lowest cost also extended the first delivery date to 2008 from 2005. So the stage was set once more.

In November 2004, the new Liberal government led by Paul Martin selected the Sikorsky *Cyclone* as the winner of the MHP. The government of Jean Chrétien had successfully avoided making a decision on a replacement for ten years. During this time, the *Sea King* helicopters experienced a series of crashes and ditches at sea – some fatal to the pilots on board. The controversial selection of the *Cyclone* and the way in which the decision had been reached predictably resulted in AWIL suing the government for over $1 billion in damages. AWIL had maintained throughout the competition that it was not possible for Sikorsky to deliver its still uncertified aircraft to Canada by the 2008 deadline. Although AWIL was fully aware that its bid would be more expensive than that of its competitor, it was certain that Sikorsky's inability to deliver would result in its being deemed non-compliant. The AWIL Statement of Claim to the Federal Court later asserted that the government chose Sikorsky despite its non-compliance because the *Cormorant* was not a politically acceptable aircraft. It alleged political favouritism and extensive errors regarding the evaluation of the bids. AWIL later settled with the government out of court in November 2007 to secure the potential for future business and because a prior Supreme Court decision undermined the strength of its case.

In January 2008, the military staff at 12 Wing in Shearwater was informed that the first *Cyclone* would not arrive until 2010 or 2011. The aircraft was still under development and was not yet certified. Sikorsky also demanded more money. DND officials had consistently held that they would avoid technology under development, especially if it was a first production run, to avoid this exact situation. The government ignored these recommendations in order to allow Sikorsky to compete. Moreover, the government refused to enforce any liquidated damages on Sikorsky for breach of contract. The contract was simply amended to allow a two-year delay and gave Sikorsky an extra $117 million to make the helicopter compliant with Canada's requirements.

On 12 March 2009, a civilian version of the future *Cyclone*, an *S-92*, crashed off the coast of Newfoundland and killed fifteen passengers and two crew who were being ferried from St. John's to an offshore oil platform. It was soon discovered that the crash was due to a loss in oil pressure and a subsequent failure in the main gearbox. The results of the investigation also revealed that Sikorsky was not technically compliant with the advanced standards of the US Federal Aviation Administration (FAA) that required the main gearbox to run for thirty minutes without oil. It was reported that the fatal helicopter crash occurred within ten minutes after the pilots reported oil pressure problems. Sikorsky is currently being sued in a Court of Common Pleas in Pennsylvania by the family members of fourteen of the deceased and the sole survivor, Robert Decker.[16]

The lawsuit states that the *S-92* is a flawed aircraft and that Sikorsky covered up its deficiencies to avoid costly redesigns. The repercussions toward the MHP program are tangible as the *Cyclone* has the same deficiencies as its civilian counterpart and must become certified before the Canadian government will accept the aircraft.

Although the first discussions of a replacement for the *Sea King* began in the 1970s, three decades of work have produced few results. After the NSA cancellation, the politics of procurement left the Canadian military without the necessary equipment to carry out its missions effectively for at least seventeen years. As this book went to publication the *Sea King* helicopters were still in operation and nobody knew when their replacements would enter service in Canada.

Acknowledgments

I had a great deal of help in the writing and preparation of this book. I must first thank Art Silverman and Will Macdonald in Ottawa. Without their help, I would not have undertaken this project at all. It would not have been possible. Much of the research on the New Shipborne Aircraft project and its eventual cancellation was retrieved as a result of their assistance and generosity. I have also relied heavily on discussions with retired *Sea King* pilots over the years. Sincere thanks go to Colonels John Orr, Larry McWha, Lee Myrhaugen, and John Cody. As they are all former commanding officers of 423 Squadron, they have been a wealth of information on the history of the *Sea King*. Indeed, they have had enough experience with the helicopter over the years! Each one has taken far too much time from their retirement to respond to my incessant questions with extensive conversations and e-mails. This information has been especially vital on a piece of equipment that is still operational, so many official files cannot be released on its deficiencies. I hope to return to the pubs in Halifax where we first met long ago during my research and return the favour. I must also thank Colonel Brian Akitt, who called me up at home one day out of the blue and directed me to all these other fine gentlemen who made this history possible. The help of Christine Dunphy, the Librarian and Archivist at the Shearwater Aviation Museum, is also much appreciated.

The work and guidance of Dr. Michael Hennessy has been vital in my understanding of ship replacement in the postwar RCN. He led me to the minutes of all the Naval Staff, Policy and Projects Co-Ordinating Committee, and Naval Board meetings where I was able to search for the information relevant to my work on helicopters. I would like to thank Mike Whitby at the Directorate of History and Heritage, Department of National Defence, for allowing me to access the extensive records of these meetings. I believe that I was the first student to go out to the new facility, and he did not even have his own photocopier yet. He later told me that he used me as an excuse to submit a proposal for one. I truly appreciated his time in talking to me about my research during copying breaks. Mike also directed me to the most important work on the beginnings of helicopter-based operations in Canada. The works of Dr. Michael Shawn Cafferky have been integral to my understanding of the early wedding of helicopters to Destroyer Escorts, and they have provided me with the necessary context to write on the history of the *Sea King* purchase. Tim Dubé also facilitated my research by allowing me access to files at Library and Archives Canada that had not yet been catalogued. Major Mike Pollard of the Canadian Forces College also proved to be an invaluable archival source.

The most important factor in my education in military history was my time at the Royal Military College of Canada. It gave me a direct understanding of my chosen field, which I have not acquired anywhere else. I will cherish going back there for the rest of my career. I must give a sincere thank-you to my supervisor, Dr. Brian McKercher. Although he was not part of this final phase of my academic journey, his mentorship during my time at the college allowed me to understand what it is to be a writer and historian. For that, I am truly grateful. I hope that my visits there will include more lively discussions in the Staff Mess.

Perhaps most importantly, I must thank Dr. David Bercuson. His tireless dedication to the field and prolific writing have been an inspiration. His high standards have only made me a better writer. Despite his obvious professionalism and discipline, his relaxed style has also made the process far more enjoyable. I hope that there are more fine cigars in our future. Thanks also go to Drs. Rob Huebert, Paul Chastko, and David Zimmerman for reading through the manuscript in its early stages. I also wish to recognize Drs. Pat Brennan and Tim Travers at the University of Calgary for their historical enthusiasm and for making me a more analytical thinker.

Many thanks go to General Paul Manson for giving me exclusive access to his personal files. I was the first to consult these volumes as they were not yet available through Library and Archives Canada. I also appreciated his reading over an early draft of the manuscript and allowing me to feel confident that I was on the right track. Lieutenant General George Macdonald and former Assistant Deputy Minister (Materiel) Ray Sturgeon also contributed to my work by giving me their valuable insights on the NSA cancellation.

The research for this book could not have been undertaken without the two years of funding from the Security and Defence Forum of the Department of National Defence. The book has been published with the help of a grant from the Canadian Federation for the Humanities and Social Sciences, through the Aid to Scholarly Publications Programme, using funds provided by the Social Sciences and Humanities Research Council of Canada. The Department of History at the University of Calgary also provided scholarships and research grants along the way. The University also gave me the opportunity to discover the joy of teaching through teaching assistantships and a position as a sessional instructor before my degree was completed. I must also thank Ron Guidinger, Malcolm Munro, Susan Premech, and the Raytheon Canada team for giving me the opportunity to see first-hand how defence procurement is done in this country.

Family members in Ottawa were also invaluable as they provided me with a place to stay during my many research trips. I was always made to feel welcome. And, of course, Jessica, who never wavered in her encouragement or patience during the final phases of the editing process. I must also thank my parents, Sharon and Al, for their enduring support. Everything that I have accomplished is because of them.

Finally, I must include a word or two on the content. This book is not a complete history of all Canadian procurement. Nor is it a complete history of the post-Second World War policies of the Canadian navy, naval aviation in Canada, or fixed-wing aircraft in the RCN. It is about the attempts to replace the *Sea King* helicopter in Canada.

Abbreviations

ACNS	Assistant Chief of the Naval Staff
ACOA	Atlantic Canada Opportunities Agency
ADM Mat	Assistant Deputy Minister (Materiel)
AERE	Aerospace Engineer
AEW	airborne early warning
AIA	Access to Information Act
AIMS	Advanced Integrated Magnetic Anomaly Detection System
AIRCOM	Air Command
AIT	Agreement on Internal Trade
ANCC	Assistant Naval Constructor-in-Chief
AOR	Auxiliary Oil Replenishment
APC	Armoured Personnel Carrier
ASW	Anti-Submarine Warfare
ASUW	anti-surface warfare
Avro	A.V. Roe
AWIL	AgustaWestland International Limited
BAMEO	Base Aircraft Maintenance and Engineering Officer
CA	Contract Authority
CAF	Canadian Armed Forces
CANDESRONS	Canadian destroyer squadrons
CANTASS	Canadian Towed Array Sonar System
CARDE	Canadian Army Research and Development Establishment
CAS	Chief of Air Staff
CASW	Canadian American Strategic Review
CCC	Canadian Commercial Corporation
CCV	Canadian Content Value
CDC	Computing Devices Canada
CED'Q	Canada Economic Development for Quebec Regions
CF	Canadian Forces
CITT	Canadian International Trade Tribunal

CMS	Chief of Maritime Staff
CNS	Chief of the Naval Staff
CO	Commanding Officer
CPF	Canadian Patrol Frigate
CSH	Canadian Search and Rescue Helicopter
CSU	Clearance for Service Use
DARMR	Director Aerospace Requirements Maritime and Rotary Wing
DCER	Documents on Canadian External Relations
DDH	helicopter-carrying destroyer
DDP	Department of Defence Production
DDPA	Deputy Director Public Affairs
DE	Destroyer Escort
DHH	Directorate of History and Heritage
DM	Deputy Minister
DMC	Defence Management Committee
DMS	Data Management System
DMS	Department of Munitions and Supply
DNAR	Director of Naval Aircraft Requirements
DND	Department of National Defence
DOI	Department of Industry
DOJ	Department of Justice
DPB	Defence Purchasing Board
EHI	European Helicopter Industries
ESM	electronic support measures
FA	Financial Authority
FAA	Federal Aviation Administration
FAR	Federal Aviation Regulations
FLIR	Forward Looking Infrared
FOD MOD	foreign object damage deflector
GDC	General Dynamics Canada
GDP	gross domestic product
GOC	General Officer Commanding
GPF	General Purpose Frigate
GPS	Global Positioning System
GST	Goods and Services Tax
HAPS	Helicopter Acoustic Processing System
HELAIRDETS	helicopter air detachments
HINPADS	Helicopter Integrated Processing and Display System
HINS	Helicopter Integrated Navigational System
HMCS	Her Majesty's Canadian Ship
HMS	Her Majesty's Ship
HOTEF	Helicopter Operational Test and Evaluation Flight/Facility
HU 21	Helicopter Utility Squadron 21

IMP	Industrial Marine Products
IRB	Industrial and Regional Benefit
ISA	international standard atmosphere
ISS	In-Service Support
LAC	Library and Archives Canada
LOI	Letter of Interest
LRPA	long-range patrol aircraft
LSO	Landing Safety Officer
LUVW	Light Utility Vehicle Wheeled
MACA	months after contract award
MAD	magnetic anomaly detection
MAG	Maritime Air Group
MARCOM	Maritime Command
MARLANT	Maritime Commander Atlantic
MARPAC	Maritime Commander Pacific
MAWS	Missile Approach Warning System
MHP	Maritime Helicopter Project
MHSOR	Maritime Helicopter Statement of Requirement
MND	Minister of National Defence
MoD	Ministry of Defence
MP	Member of Parliament
MRG	Management Review Group
MTBF	mean time between failure
NATO	North Atlantic Treaty Organization
NBCD	nuclear, biological, and chemical defence
NCC	Naval Constructor-in-Chief
NDHQ	National Defence Headquarters
NDP	New Democratic Party
NFA	New Fighter Aircraft
NORAD	North American Air Defence Command
NPCC	Naval Policy Co-Ordinating Committee
NSA	New Shipborne Aircraft
NSH	New Search and Rescue Helicopter
NSS	Naval Secret Staff
NVG	Night Vision Goggles
PM	Prime Minister
PMO	Project Management Office
PPCC	Policy and Projects Coordinating Committee
PPP	program planning proposal
PRM	progress review meeting
PWGSC	Public Works and Government Services Canada
RA	Requisitioning Authority
RAF	Royal Air Force

RAST	recovery, assist, secure, and traverse
RCAF	Royal Canadian Air Force
RCN	Royal Canadian Navy
RCNAS	Royal Canadian Naval Air Service
RFP	Request for Proposal
RMC	Royal Military College
RNAS	Royal Naval Air Service
RN	Royal Navy
RS	Requirement Specifications
SAM	surface-to-air missile
SAR	Search and Rescue
SCONDVA	Standing Committee on National Defence and Veterans Affairs
SENSO	Sensor Systems Operator
SKIP	*Sea King* Improvement Program
SKR	*Sea King* Replacement Program
SMP	Standard Military Pattern
SOI	Solicitation of Interest
SOR	Statement of Requirement
SOSUS	Sound Surveillance Systems
SPAC	Senior Project Advisory Committee
TA	Technical Authority
TACAN/DME	Tactical Air Navigation/Distance Measuring Equipment
TACCO	Tactical Co-Ordinator
TRUMP	Tribal Class Upgrade and Modernization Project
TSB	Transportation Safety Board of Canada
UN	United Nations
USN	United States Navy
USS	United States Ship
VCDS	Vice Chief of the Defence Staff
VCNS	Vice Chief of the Naval Staff
VDS	Variable Depth Sonar
V/UHF	Very High/Ultra High Frequency
VISIT	Vertical Insertion Search and Inspection Team
VX 10	Experimentation Squadron Ten
WD	Western Economic Diversification

The Politics of Procurement

A formation of three CF-18 *Hornet* Aircraft on 26 October 1986.

Source: CF photo by Sgt Boies. *Courtesy of National Defence. Reproduced with the permission of the Minister of Public Works and Government Services, 2009*

Introduction: The Canadian Defence Procurement System

Any book on defence procurement should start with how the system is supposed to work. This is not an easy task. As Janet Thorsteinson, vice president of government relations at the Canadian Association of Defence and Security Industries, has stated, "Canada's legal and policy procurement framework is more complex than that of any other nation."[1] Although there is a standard procurement outline, there are often internal clauses that make each acquisition unique. In fact, the process is often a "moving target."[2] A brief history of the process, however, and a description of some of the major changes within it will serve the reader well regarding what each stage *could* look like in any Canadian military acquisition. Although the steps have been given different names over time, the procurement process has essentially included a definition of military requirement; validation of the requirement; government approval of the project, including funding; creation of an official Statement of Requirement (SOR); selection of a procurement strategy; bid solicitation and source selection; negotiation and award of contract, with the possible inclusion of another for long term In-Service Support (ISS); administration of the contract to purchase the piece of equipment decided on; and finally delivery of the product.[3] Although most of these phases were present throughout the period of this study, the time that it took to complete each phase continually increased. Further judgment on the process occurs in the following chapters with the history of the ongoing attempts to replace the *Sea King* helicopter.

The First World War revealed the deficiencies of the procurement system in Canada.[4] In short, there was no effective coordination in dealing with purchases made in Canada by the Allies. Domestic acquisitions were also done ad hoc by each military service (army, air force, and navy), which procured independently of one another, and there were often significant variations in the prices paid by each.[5] Profiteering was also a major concern. Investigations into the origins of these problems carried on throughout the interwar period in an attempt to avoid a repetition of this experience in any future war. Although the Department of National Defence (DND) established the Navy, Army, and Air Supply Committee in 1936 under the chairmanship of the Master General of the Ordnance, military procurement continued to be conducted without proper guidelines. This was revealed with the Bren gun contract of 1938. As early as the summer of 1936, the DND came to the conclusion that it would be necessary to arm the Canadian Forces (CF) with the Czech-designed Bren light machine gun.[6] Although an interdepartmental committee was appointed by Prime Minister Mackenzie King in January 1937 to report on the control of profits on armament contracts, the government was still inexperienced with the intricacies of contracting and weapons development.[7]

On 31 March 1938, a contract for the production of 7,000 Bren guns was signed with the John Inglis Company of Toronto. The total order became 12,000 when the British War Office also agreed to purchase 5,000 of the guns. Although this meant that Canada was successfully getting involved in its own defence production, the contract was quickly criticized since no company besides John Inglis had been given an opportunity to tender for it.[8] The contract illustrated the struggle between private industry and the Canadian government over who should take primary responsibility for the construction of military materiel. A Royal Commission was appointed to investigate the situation and concluded that the Interdepartmental Committee on Profit Control provided inadequate protection against profiteering. It recommended that any negotiations between the government and private manufacturers regarding armament contracts should be put into the hands of an expert advisory group of competent businessmen.[9] The government then concluded that a centralized procurement agency, composed of civilians with experience in purchasing, was needed to deal with both Canadian military requirements and anticipated export demands. It was thought that this agency would lead to better economy and administrative efficiency. Accordingly, the Defence Purchases, Profits Control, and Financing Act was passed in 1939 and authorized the appointment of the Defence Purchasing Board (DPB), which began operations in July 1939. Although there had been a wide breach between the spheres of civilian and soldier prior to the First World War, the vast nature of military expenditures and budgetary challenges associated with defence preparedness in the age of total industrial war demanded a new form of co-operation.[10]

As a result of the increased equipment demands of the Second World War, the DPB was replaced by the War Supply Board on 1 November 1939, and by the spring of 1940 a full department had been established to handle military procurement. The Department of Munitions and Supply (DMS) was assigned far-reaching control and operated under special emergency legislation; it was given the power to buy, sell, ration, allocate, or fix the prices of essential supplies and to establish priorities if necessary. In brief, this department was empowered to direct and control war production in any way necessary for the furtherance of the war effort. The DMS handled procurement for the Canadian Forces and also for the United Kingdom, the United States, and other allies to the extent that these countries purchased in Canada. Due to the distinctiveness of military purchases, the DMS also had an independent contract authority entitled the Contracts Authorization Division. Non-military supplies were handled by the Canadian Export Board, which had been created within the Department of Trade and Commerce on 31 January 1940.

Military and civilian planning and production were combined on 18 December 1945, when the Department of Munitions and Supply and the recently formed Department of Reconstruction were amalgamated as the Department of Reconstruction and Supply. The new ministry assumed the duties of its predecessors, which were to construct and maintain buildings and bases and to acquire anything deemed necessary for war or reconstruction, while retaining investigative and enforcement powers to ensure compliance with contracts.[11] After the war was over, the emergency powers of this new department were no longer necessary, and military procurement was passed to the Minister of Trade and Commerce in February 1947. From that point on, military acquisitions went through the minister, who used a civilian purchasing agency, the Canadian Commercial Corporation (CCC), which had been established during the war to handle non-military purchasing. It performed the same general functions for the services as had been performed by the DMS: namely, receiving from the services details of their requirements, canvassing the market to determine the best source of supply, awarding the contracts, and following up on deliveries. It did not assume responsibility for inspection of delivered goods, nor did it pay the suppliers; these two functions were the responsibility of the DND.[12]

To meet the needs of an expanded defence program after the outbreak of the Korean War, a separate Department of Defence Production (DDP) was established in April 1951 and took over the role of purchasing agent for the government. The essentials of military procurement, developed during the Second World War, were carried over into the new department. The DDP was only responsible for defence acquisitions; it inspected, constructed, and acquired defence projects and supplies on behalf of the DND while mobilizing, conserving, and coordinating all economic and industrial facilities necessary for military and civil defence. Several crown corporations were immediately transferred to the DDP: Canadian Arsenals Limited, Crown

Assets Disposal Corporation, Defence Construction Limited, Polymer Corporation Limited, Eldorado Mining and Refining Limited, Northern Transportation Company Limited, and the CCC. By these means, the government again supervised the manufacture and sale of essential commodities, such as steel and uranium, as it had done during the Second World War. The Minister of Defence Production retained the right to create crown corporations, fix prices, limit profits, and compel services deemed essential for Canada's defence.[13]

Once a requirement had received internal approval within the DND, a purchasing action was initiated by raising a requisition or contract demand. After that, a contract demand was submitted to the DDP, and purchase negotiations could be started with prospective suppliers. Acting as the CCC's agent, the ministry purchased defence supplies from Canadian companies on behalf of foreign governments. As part of a government-wide initiative, in May 1960 the DDP established an Emergency Supply Planning Branch to plan for the immediate creation of a War Supplies Agency in the event of a nuclear attack.[14] Since the first decision to use civilians for military purchasing, there had always been the issue of them being handicapped by their unfamiliarity with military requirements and the attendant jargon. This problem was circumvented to a large extent by the secondment of military personnel from the three services to assist the civilians within the DDP in dealing with highly technical and complicated equipment.[15]

Further change came as the Liberal government of Lester B. Pearson reformed the federal civil service. On 25 July 1963, the duties of the DDP were transferred to a new Department of Industry (DOI). On 4 September 1963, following the Glassco Royal Commission on Government Organization, a central purchasing and supply agency for all civilian departments and agencies – other than commercially oriented crown corporations – was formed when the DOI established Canadian Government Purchasing, Supply, and Repair Services. That year an International Programs Branch was established to "guide and co-ordinate all aspects of the Department's international defence cooperation and export programs," including marketing and production sharing with England, Europe, and the United States.[16] These DOI branches administered the Canada-United States Defence Production Sharing Agreement and appointed overseas attachés to coordinate NATO defence production. As Cold War tensions finally abated, the DDP was replaced by the new Ministry of Supply and Services on 1 April 1969 under terms of the Government Organization Act.

After Trudeau and the Liberals came to power in 1968, a Management Review Group (MRG) was appointed by Minister of National Defence Donald Macdonald to examine all aspects of the management of the DND. They finally concluded their work in 1972, and the report changed the entire administrative structure of the DND and the Canadian Forces, which had been unified in 1968.[17] The report recommended that all military research, engineering, and procurement be consolidated under one Assistant Deputy Minister, Materiel (ADM Mat), within the department. Whoever filled the position was to be a civilian with experience in industry. The report maintained that procurement of military equipment should fall solely within the DND due to its complexity and cost, and these specific items would be handled by the new ADM instead of through the Department of Supply and Services. The MRG believed that this change would provide a focal point of accountability, much as the government had when it created the original civilian agency, the Defence Purchasing Board, in 1939. In 1972, the military and civilian elements of the DND and Canadian Forces Headquarters were integrated into a new Canadian National Defence Headquarters (NDHQ). The new position of ADM Mat was created within the new organizational structure. The ADM Mat became responsible for long-term equipment planning and logistics, but the position was also to be responsible for the contracts themselves and was designed to facilitate effective procurement.[18]

The next major change was the creation of the Department of Public Works and Government Services Canada (PWGSC) in June 1993 when four departments – Public Works Canada, the Translation Bureau, Supply and Services Canada, and the Government Telecommunications Agency – were combined.[19] In 1996, legislation confirming the merger and establishing the new department was enacted. The department has operated as a common service agency for the Government of Canada ever since. At the time of publication, it was responsible for more than 100 federal departments and agencies each year and had to cover thousands of different purchasing categories. With its creation, there were no longer any special guidelines that differentiated military acquisitions from other government purchases. As A. Crosby has asserted, "that is quite different from what you'll find in many other countries around the world, such as the United States, Great Britain, and so on, where their department of defence has specific procurement authorities and specific rules."[20]

Although the procurement process has continued to evolve since the creation of PWGSC, Canadian equipment procurement has been composed of key elements and has followed a general pattern for major equipment projects, which are those over $100 million.[21] Before receiving funding, the individual military services have always been required to internally generate their demands before seeking formal approval from their chiefs of staff. The process has been different for each service and has altered over time. The navy, responsible for helicopter procurement during the 1950s, for example, had the Policy and Projects Co-Ordinating Committee, which made recommendations to the Canadian Naval Staff, which met and made further recommendations to the Naval Board. The board then passed on their recommendations on necessary equipment to the chiefs of staff, who then passed on their conclusions to the Minister of National Defence through the Deputy Minister.[22] Regardless of how each service generated its proposed requirements, these have always been statements of capability deficiency. A preliminary list of potential solutions is then developed, along with initial cost estimates. The project is then commonly entered into a long-term capital plan, which signifies an agreement at the DND level to address the deficiency. Responsibility is then likely assigned to an environmental commander – land, sea, or air – to proceed with the next phases of the project.

The next common phase is the Project Development Phase. The environmental commander chairs a senior departmental committee and approves an overall plan for the project, which outlines what has to be done, who the major people are, and the organizations that will be involved. A project manager is then named. As no funding has been approved by this stage, all staff are internal, and no independent project office has yet been established. Option analysis and risk assessment are carried out along with a refinement of cost estimates against the various options.

A SOR is then compiled and reviewed by the DND's senior management oversight committee. The project profile and risk assessment are then prepared and reviewed with the Treasury Board, composed of the Minister of Finance and other selected cabinet ministers. An initial procurement strategy is developed and reviewed by the interdepartmental Project Advisory Committee for major projects. All this leads to finalizing what is called the Preliminary Project Approval. It is then reviewed by those involved within the DND and then submitted to the Deputy Minister and then to the Minister of National Defence. If approved by the minister, the project plan is then taken to cabinet. If it passes this phase, cabinet submits the project to the Treasury Board for approval of preliminary funding. Essentially, this approval allows further definition work if it is required and sanctions in principle the project based on the rough estimates that have been developed at this stage.

With monetary approval to proceed, the project can enter into a Project Definition Phase. This phase is often long and complicated. In essence, it is the phase during which the desired end product is described in detail. This description includes the minimum standard of operational performance required by the Canadian Forces. It also includes analysis of existing and available technology, often referred to as items "on the shelf"; evaluation of the desirability of a developmental "off the shelf" purchase; budget projections; logistical support requirements; and industrial benefits defined by Industry Canada. Costs, schedules, and risk parameters are finalized along with the development of the SOR. The SOR is reviewed and approved once again by the DND senior management oversight committee, and the preferred option is then selected. The project profile and risk assessment are updated based on this preferred option and again reviewed with the Treasury Board Secretariat. At that point, more discussions on the risk of the project occur.

For major crown projects, there is automatically an interdepartmental Project Management Office, along with the official Senior Project Advisory Committee (SPAC), established in accordance with Treasury Board guidelines. The interdepartmental Project Management Office is comprised of a team of professionals assembled to help make key decisions on the procurement strategy. It includes PWGSC. Other key interdepartmental members may include the Privy Council, Finance Canada, Industry Canada, Human and Resources Development Canada, Indian and Northern Affairs, Environment Canada, and the National Research Council. The Department of Foreign Affairs and International Trade monitors defence import and export policy. The number of organizations or departments involved has grown and changed over time. What is important to understand is that only the DND is truly worried about actual military capability, whereas the rest are concerned more with things such as regional benefits, national job creation, and taxes.

Canadian defence procurement often includes the need to have Canadian Content Value (CCV). This means that the company awarded a contract must perform part of the work in Canada and use Canadian materials. Although Industrial and Regional Benefits (IRBs) have been a focal part of Canadian procurement throughout its history, particularly in the 1970s, this aspect of defence acquisition was passed by cabinet and became official Canadian policy in 1986.[23] The IRB policy provides the framework for using major federal government procurements to support prolonged industrial and regional development. IRBs are business compensation activities undertaken by the prime contractor in Canada as a result of successfully bidding on a Canadian defence procurement. IRBs can contribute widely to the national economy, including helping small business and improving the national balance of trade. They are important contractual commitments and usually involve highly sophisticated technology that is designed to bring lasting value to the Canadian economy, without the government being responsible for production. As Stephen Martin has explained:

> While both co-production and licensed production provide employment and technological benefits for the domestic economy, the establishment of an indigenous production line is a costly business ... purchasers have sought a less costly form of procurement which still generates work for the domestic economy. Typically, this obliges the foreign vendor and its sub-contractors to buy goods and services over and above what it would have bought from

> firms in the purchaser's economy. This offset is usually some percentage of the contract price and a time period is often set for its fulfillment.[24]

Although Industry Canada has the lead responsibility for the IRB program and for establishing what are acceptable IRB credits for a company, the Atlantic Canada Opportunities Agency (ACOA), Western Economic Diversification (WD), and Canada Economic Development for Quebec Regions (CED'Q) have representation on the SPAC. The IRB policy is mandatory for major crown projects, and prime contractors must commit to achieving benefits in Canada equal to 100 percent of the contract value. The company must identify 60 percent of these benefits in its bid.[25] These reciprocal investments can either be direct or indirect. A transaction is considered direct if the business activities provide goods, services and/or long-term service support directly for the items being procured by the government. Business activity that is not directly related to the procured items is considered indirect and can be accomplished through investments in areas such as research, education, agricultural commodities, technology cooperation, or raw materials. If a procurement is deemed able to give industrial benefits to Canada, an IRB program is created through Industry Canada, in co-operation with the client department, the DND, PWGSC, and the regional agencies. Clearly, navigation of this bureaucracy takes extended periods of time.

It is at this stage that the final procurement strategy is chosen. It could be based, for example, on the lowest cost or the best value, the latter being a combination of price and quality. It is also determined whether it will be a single contract or split into two to distribute the IRBs. It is often at this time that the government is accused of making a final decision at odds with those who made recommendations based on military capability needed by the CF. The interdepartmental SPAC then reviews the final procurement strategy; if they agree to it, Effective Project Approval can be finalized. The process then needs to be approved by the Minister of National Defence, cabinet, and the Treasury Board. This approval, when it is complete, gives the government expenditure authority for implementation of the project.

The project office then approves the final SOR. Preparation and approval of the SOR are logically preceded by an operational analysis to explain where and when the platform will be employed, against which threats, and what it is expected to accomplish generally. The military requirements set out in the SOR are described in the technical language of industry so that potential suppliers can make a fully informed decision on whether they can comply with the specifications and bid on the contract. This is also done so that government contract authorities can establish the criteria required to evaluate the bids.

This brings us to what is commonly called the Implementation Phase. Prior to the formal Request for Proposal (RFP) released to possible bidders for the contract, the federal government could use a number of methods to interact with industry. A Letter of Interest (LOI) can be released to possible contenders for the contract. It is meant to eliminate unsuitable suppliers at an early stage, thereby saving time and money for both business and government.[26] Most commonly, the contract authority – PWGSC – issues a Solicitation of Interest (SOI) to industry on behalf of the client, the DND. Considering the highly technical nature of defence products, usually only a small number of companies will reply. On some occasions, the RFP may be preceded by a prequalification phase, and only the potential bidders found to be acceptable will receive the RFP. This was the case with the Maritime Helicopter Project (MHP). Part of this prequalification screening involves the Project Management Office meeting with the bidders to discuss the project. This meeting ensures that suppliers are fully aware of the requirements and gives them an opportunity to shape their bids before the formal RFP is issued. This process also includes meeting with Industry Canada and the regional agencies to discuss any potential IRBs. The bidders can then start discussions with other Canadian companies if domestic involvement is a necessary part of the contract; time is needed to negotiate with them, find an affordable solution, and develop long-range business plans.

Finally, the project office issues the RFP. It is usually posted on MERX, the Canadian electronic tendering service, and becomes publicly available; even companies that did not respond to the LOI are usually eligible to respond to the RFP. The project office then evaluates the bids received and selects a winning bid based on a number of important elements. They may include price, value, technical merit, delivery, contract terms and conditions, risk, and IRBs. The project team also evaluates specific proposals such as the Project Management Plan, the System Engineering Management Plan, and the long-term In-Service Support (ISS) Management Plan.

The SPAC then makes a recommendation to the ministers involved on which proposal should be selected for contract award, and a winner is announced. The winning bidder is then invited to enter into contract negotiations with the government. These negotiations are done through the contract authority (CA) from PWGSC, the requisitioning authority (RA) from the DND, which is the liaison between PWGSC and the DND, and the company chosen to carry out the project. There is also the technical authority (TA), which works for the DND and is responsible for technical compliance of the contract, and the financial authority (FA) from the Treasury Board to oversee the exchange of capital. If there is an IRB component, the IRB manager from Industry Canada negotiates with the company directly to ensure that it will fulfill the IRB portion of the future contract. The IRB manager will then decide on compliance and communicate the decision directly to PWGSC.

There are times when the chosen contractor is funded to further define its solution to the stated requirement and to provide to the government a fully budgeted and substantiated proposal. This was what happened with the contract for the New Shipborne Aircraft (NSA) program signed with European Helicopter Industries (EHI) to replace the *Sea King* helicopters in 1988. Once all the terms and conditions have been met and all parties have come to an agreement, the Treasury Board must give its final consent, and contract approval is then obtained with the consent of cabinet. The final contract is then signed. From that point on, the project office is responsible for managing the delivery of the equipment, which includes ensuring contract compliance; monitoring progress; authorizing payment; managing amendments; and ensuring that, when the equipment is delivered, there are already trained personnel ready to put it into service with the other armed forces. Progress review meetings (PRMs) are usually conducted annually to discuss any necessary amendments to the contract and overall performance of the contractor. Perhaps most important to the contract, the company is responsible for delivering the product; liquidated damages are assessed for any non-compliance. The entire process is then subject to audit by the Auditor General.

As the reader will discover, the process is even more complicated than it seems. Each step within this labyrinth has provided the opportunity for budgetary and scheduling pitfalls. There is no better way to reveal these systemic challenges than to tell the story of the *Sea King* helicopter in Canada.

Left side view of a CF-105 Jet Fighter Aircraft Avro *Arrow* in flight.

Source: CF photo. *Courtesy of National Defence. Reproduced with the permission of the Minister of Public Works and Government Services, 2009*

1
Procurement in Canada: A Brief History

> Canadian equipment purchases had always involved politics, right back to the 1880s's decision to dress the militia in high cost, low quality Canadian made uniforms in deference to Sir John A. Macdonald's National Policy.
>
> – Desmond Morton, *Understanding Canadian Defence*

> The matter of militia equipment leads us to consideration of perhaps the most fundamental, and at the same time the most difficult and controversial problem which arose in connection with the new Canadian defence programme: that of the procurement of armament and equipment. Apart from the great temporary development which took place during the Great War, Canada has never possessed an armament industry of any importance; even at that period, we have noted, no *weapons* were manufactured except Ross rifles. In these circumstances, she was bound to find herself in difficulties when the need suddenly arose for modernizing her fighting services.
>
> – C.P. Stacey, discussing the lack of preparation for the Second World War

For the majority of its history, Canada has been incapable or unwilling to properly equip its military. Failure to design and produce the necessary materiel to support the Canadian Forces (CF) domestically consistently led the government and the Department of National Defence (DND) to search for foreign alternatives. These purchases of foreign military equipment were often necessary due to the limited size of Canadian industry and its inability to compete in international markets – this is often called the economies of scale.

This reliance on foreign sources has frequently hampered the scheduling of Canadian military procurement and created various other problems regarding the introduction of weapons platforms. Although the necessity of equipping a military for operations is equally important to all other factors involved in preparing a national defence force, weapons and equipment procurement in Canada has generally been inefficient.[1] Scandalous acts committed for political gain are not unexpected. It is difficult to assess this fully, however, as the topic has received scant attention from Canadian military historians. As David Bercuson has asserted, "in any well-stocked bookstore today there will be tomes on great military leaders, decisive battles, the evolution of strategy and tactics, intelligence, the art of war, military leadership, even supplies, logistics, and communications. But nothing on procurement."[2]

The absence of literature on procurement is most notable during the period before Canada's entry into the Great War. There are many publications on the early defence policies, attempted reforms, and personalities involved in Canada's military development before 1914.[3] But there is still extremely little on how the Canadian military of that time equipped itself. One reason is that the political parrying between the General Officer Commanding (GOC) and the government's representative, the Minister of Militia, is truly a fascinating story of political patronage, corruption, and lassitude. The other reason is that there is very little to tell; Canada's early militia was thinly armed, and the weaponry used was rarely kept current with other international standards, especially regarding the soldier's basic rifle. The reason, in the early years of the nation, was twofold: first, Canadian governments believed that the British would always be there to save them in the face of an emergency, specifically against the Americans; second, these governments had concluded that there was little tangible threat from the Americans themselves. In short, there was no impetus for Canada's leaders to invest time, energy, and, most importantly, money into the Canadian militia. As C.P. Stacey has asserted, "[Canadian] history is full of warlike episodes, and they have proved on many an occasion that they can be skillful and determined fighters; but few nations have shown more profound antipathy to the idea of military preparations in time of peace, or less interest in military affairs generally except in moments of emergency."[4] Nowhere has this lack of preparation been more obvious in Canada than in the field of weapons procurement.

Canada has been buying weapons of war since before Confederation in 1867; unfortunately, only certain areas are covered in the literature, and the history of Canadian acquisition policies has remained untold. There exists extensive commentary on the Canadian Ross rifle and the munitions industry of the Great War.[5] Most authors, however, have focused on the industrial defence effort of the Second World War and the postwar period, particularly the Avro *Arrow* and the bilateral Defence Production Sharing Agreements between Canada and the United States.[6] As one of the few authors on the subject, Dan Middlemiss, has written, "notwithstanding the availability of many useful procurement case studies, what is lacking is a general overview of weapons acquisition in Canada."[7] This trend likely exists because of the perception that, as one historian has put it, "with the exception of a Government Factory established in Quebec City in 1882, and the Ross Rifle fiasco of 1904-15, the Canadian government made no attempt to establish an armaments industry or even to develop an industrial preparedness policy until just prior to the Second World War."[8] These efforts offer an incomplete understanding of Canadian procurement history. Even the latter study, which claims to be a history of Canada's defence industrial policy, begins in 1935.

Although the present work will not remedy the lack of a comprehensive narrative on procurement history, a few incidents in Canadian history will be highlighted to demonstrate some dominant themes. These trends began even before Canada was a country. In 1862, John A. Macdonald, the first administrator of the Militia Affairs portfolio for British North America, received the first indication that the electorate was not willing to invest in its own defence. The American Civil War and the subsequent Trent Affair of 1861 – where the federal navy seized Confederate envoys aboard a British vessel – had heightened tensions between the United States and Britain. Although direct conflict was averted, the British had begun military mobilization, and it resulted in increased calls for defence improvements in British North America. A Militia Commission was formed in Canada in 1862. When its recommendations came back within the year, Macdonald used them as the basis for introducing a militia bill. The commission reported the need for a trained force of 50,000 and a reserve of the same size. It quickly became clear to the opposition that such a force would cost approximately $500,000 and that no such investment was possible; instead of proposing amendments, the opposition saw it as a giant target and attacked it absolutely. The Cartier–Macdonald ministry was in a weak position at the time, and with Macdonald drinking heavily the bill was poorly defended; it was subsequently rejected, and the Cartier–Macdonald ministry resigned the next day.[9] The bill called for an expenditure that was seen by the people as too large for a small colony. From the perspective of the British, its rejection meant that their colony had no intention of defending itself. This example also made politicians in British North America keenly aware that defence expenditures were politically dangerous.

In the 1880s, the government focused on building a public arsenal system to produce all war stuffs for the Canadian militia. The Canadians, for their part, were also more forthcoming than usual on defence as there were riots in Quebec in 1878 and talk of a resurgence of the Irish Fenians in the United

States.[10] The arsenal would be government owned because there was no company willing to undertake the scheme. When the Conservative administration approved the project at the Citadel in Quebec City in 1880, however, it was decided that only ammunition would be produced. Guns and other necessities would, for the most part, still come from England. Cost was always the biggest factor in further developing the arsenal; it remained little more than an assembly plant for expensive imported British materials to create a small number of cartridges. As one author put it, "Canadians, having become accustomed to bearing no responsibility, and little of the cost of their own defence ... in view of the ever decreasing external threat from the USA, thought that any expense, however small, was a waste of money."[11] In addition, it never became a full arsenal system with competing contractors because it was politically healthier for the government to reserve the small number of contracts for industrial friends. Politicians were certainly not going to turn these patronage possibilities over to the military for the sake of development and efficiency; these deals were reserved for loyal party followers who happened to be interested in the military.[12] The nature of early Canadian equipment procurement was, thus, that of political favouritism.

Weaknesses in Canadian equipment during the South African War of 1899-1902 placed pressure on the government to acquire a new rifle for the militia. For the first time, thoughts that concerned defence materialized into serious interest in weapons procurement. In 1902, Minister of Militia Frederick Borden investigated whether British weapon designers would come to Canada and build a rifle. It was believed that it would be more practical to have the guns built domestically to be able to produce more in a time of crisis. After this proposal failed to create any interest, Borden decided that Canada would adopt its own rifle made in Canada and designed by a local entrepreneur, Sir Charles Ross.[13] The government subsidized production of the Ross rifle, and it became one of Canada's first weapons purchased primarily for political reasons. As Ronald Haycock has pointed out,

> Political manipulation of procurement was rife because acquisitions came under the civilian sphere of the defence department, where they were controlled by the Deputy Minister. The military had no input into this area until well after the turn of the century, and even then acquisition would remain more a political process than a military one. Most often in the post-Confederation decades, the civilian contractors had to be of an acceptable political persuasion, as the Canadian Militia was constituent based and highly politicized, and because few cared about defence.[14]

Ross was subsequently given a contract in 1902 and a twenty-five-acre site to build a factory in Prime Minister Laurier's constituency near Quebec City. Ross paid a dollar a year for rent. But between 1904 and 1907, he failed to produce the number of rifles stipulated in the contract. He was still being paid in full, however, due to his political friendships.[15]

At the Imperial Conference of 1909 on the Naval and Military Defence of the Empire, Sir Richard B. Haldane, the British Secretary of State for War, inquired, "are the Dominions prepared to adopt as far as possible imperial patterns of arms, equipment and stores?"[16] Borden, still the Canadian Minister of Militia, responded that he agreed that all arms should be identical, and the only reason that Canada had a different service rifle than the British Lee Enfield was that he had been unable to convince English manufacturers to establish factories in Canada. He assured Haldane that Canada took "good care to secure a rifle" that used "the same ammunition as the Service rifle of the Imperial army," and therefore there would be little difficulty in coordination. Borden felt certain that Canada was in a propitious position to unite with the imperial military, and he confidently stated that using British models was simply common sense and that it took nothing away from local autonomy. The dedication to being independent militarily was, therefore, not adamant, and the Canadians were very receptive to the imperial military pattern regarding their equipment. If they could procure British style kit, they did.

The rest of the story of the Ross rifle is well documented elsewhere.[17] What needs to be understood is that Canada had attempted to build its own weapon and that it was universally determined to be a failure on the battlefield. The Ross was an excellent target rifle, but it was deficient as a service weapon. It was unable to fire rapidly without overheating and seizing up. On the battlefields of Ypres, it was reported that some 3,000 men cast aside their Canadian rifles, most jammed with mud, and armed themselves with British weapons.[18] In a strange twist that showed the politics of the matter, Ross eventually sued the government for $3 million and was given $2 million in an out-of-court settlement after the Deputy Minister of Justice advised the government that Ross had a good case based on the vagueness of his contract.[19]

The Canadian service rifle was not the only piece of kit considered a failure. The Canadian contingent was originally to be supplied with its own boots and greatcoats. But because of profiteering, much of the manufactured equipment was of poor quality. Two million boots ordered by the War Office were useless as they had been made largely of cardboard and fell apart in the rain.[20] Canadian coats were too thin and inferior for British use. It also became known that the Canadian Service Wagon was rejected as it had a turning radius too large for English and European roads.[21] The McAdam shovel-shield carried by all Canadian soldiers was useless for digging and cutting wire, and they were sold for scrap in 1917.[22] The dominion's primary contribution was artillery shell production. The Shell Committee was created on 6 September 1914. But due to political patronage, profiteering, production failures, and extensive delays in shell delivery, it was largely a failure.[23]

Although shell production was revamped and improved over the course of the war, the Canadian attempt to deviate from British weapons and equipment models did not succeed.

Sir John Stevens, the British Director of Equipment and Ordnance Stores, stated in 1917 that, after the war,

> The general and guiding principle which it is desired to submit is that appropriate steps should be taken to combine the experience of the whole of the Imperial forces in the selection of the most suitable designs of equipment and clothing, and that the designs thus selected should be adopted throughout the Empire with such variations alone as are necessitated ... by differences in climate and other local conditions.[24]

All the dominions agreed.

By the end of the First World War, Canada's confidence concerning domestic military industry was irreparably shaken due to the failure of its indigenous military equipment.[25] Moreover, Canada was financially weakened by the war. It became more expedient, therefore, for the Canadian government to use British models than to pay for independent research and development for its own specific equipment needs. Notwithstanding a constant Canadian quest for independence – best illustrated in the government's insistence on a separate signature on the Treaty of Versailles and its independent membership in the League of Nations – Canada still deemed it expedient to use British military prototypes. As a result, equipment was dictated to Canada because it did not design patterns expressly for its needs; the British were not building for Canadian requirements. Moreover, because of time constraints, Canada often did not have the time to recommend changes to British designs, as such changes would have meant inevitable delays in procurement. Because the Canadian equipment base was already quite sparse, the government simply could not wait for suitable machines to be designed. This meant that Canada was constantly seeking approval for modifications to its own equipment, which often were not possible, due partly to the difficulty in exchanging blueprints. Even something as simple as an aircraft starter had to be ordered from England for £250. When the manufacture of the starter "had ceased to exist and no drawings were available," Canada was simply out of luck. The Air Board decreed that "if we had a complete machine we could use it as a pattern and have more made in this country."[26] But they did not. So, in addition to determining which parts were altered, Britain decided when to do so; to keep its planes air worthy, Canada had to follow course. The historic lack of investment in the defence industry has often hindered Canadian military capability, and foreign equipment purchases have placed the Canadian Forces at the mercy of foreign technologies, processes, and political decisions.

Canada even refused to make socks for its soldiers. An Air Board meeting declared in 1922 that "the purchase of waders brogues and socks in England for the Operations Branch was approved at a cost of approximately $800. The Inspector General asked that a submission be made for the Minister's sanction."[27] This was myopic in terms of long-term military planning because it did not allow for even the most basic of military kit to be produced in Canada. In the event of a crisis, the Canadian military could easily lose its supplier. This was common practice in postwar Canada. The Canadian government did not desire any freedom concerning equipment designs, and this trend continued at the Imperial Conference of 1926, where imperial standardization again became absolute.[28]

Although equipment designs were dictated by the Empire, Canada did attempt to manufacture certain weapons domestically during the 1930s. And the contract to build the Bren light machine guns is another example of the political nature of procurement in Canada.[29] The primary result of the supposed scandal was that the Canadian government brought in legislation to establish a Defence Purchasing Board (DPB).[30] Although the DPB was a good idea, C.P. Stacey has astutely asserted that there was nothing wrong with purchasing the gun from a sole source as it had been decided that it was necessary from the standpoint of military capability. He correctly wrote that "this proved to be one more case where political considerations took precedence over military expediency with unfortunate results."[31] The Canadian government, however, was not yet competent in the direct involvement of arms production in the 1930s.

The Second World War forced the Canadian government to invest in its defence industry, and domestic production effectively supplemented the traditional British source. Britain was economically devastated by the war and could not help to arm Canada. Domestic industry subsequently became responsible for *Anson* and *Harvard* training planes, *Mosquito* fighter bombers, *Hurricane* fighters, and *Lancaster* heavy bombers. None of the planes was designed in Canada, and no aircraft engine was made domestically.[32] The engines either had to be imported or installed elsewhere.

Canada still did not have trained engineers for its own defence design and construction, and Canadian industry continued to use foreign models. Examples from Britain include the twenty-five-pounder field guns, 3.7-inch anti-aircraft guns, two-pounder anti-tank guns, and Boys anti-tank rifles.[33] During the war, 815,729 transport vehicles and trucks were built in Canada, and they represented one of its most distinguished industrial achievements; however, this situation was facilitated by American dominance over the Canadian automobile industry.[34]

Canadian industry also built corvettes, frigates, and *Tribal* class destroyers for the Royal Canadian Navy (RCN).[35] Although four were ordered in Canada in 1941, the *Tribal* destroyers were not completed until the end of the war due to a higher priority being given to other tasks. Canadian Minister of National Defence J.L. Ralston later claimed that the British Admiralty had been unwilling to lend naval personnel to assist in destroyer construction in Canada.[36] Indeed, the lack of weapons specialists in Canada created many difficulties for the RCN. This problem was most notable in the field of Anti-Submarine Warfare (ASW), which Canada focused on during the war. As W.G.D. Lund has argued, "it was through its commitment to anti-submarine warfare that Canada was able to gain some measure of control in the Battle of the Atlantic."[37]

The RCN played a role second only to the Royal Navy (RN) in the protection of trade routes in the North Atlantic, and this endeavour created a need for highly developed ASW technology. The RCN had to keep pace with the science of radar, asdic, high-frequency direction finding, and offensive anti-submarine weapons, such as the Hedgehog ahead-throwing mortar. At the outbreak of the war, however, there was not one technical or scientific adviser in Naval Service Headquarters in Ottawa. All of the RCN's weaponry came from Britain, and after the war started and British supplies disappeared, Canada was on its own. The attempt to design and procure advanced technology necessary for the success of the RCN in the North Atlantic has been described as "a national failure."[38] David Zimmerman has written that,

> Even in weapon design their efforts were a failure because of the dissimilar priorities of the institutions involved and the inevitable conflict that developed between them. The National Research Council of Canada, the supreme wartime scientific agency, and Naval Service Headquarters did not succeed in resolving their difficulties, the effects of which on the anti-submarine campaign were profound as RCN escorts went to sea with inferior, outdated, or unusable equipment.

One exception regarding weapons design in Canada was the *Ram* tank. Although it was based on the American *M-3* medium tank, Canadian engineers believed the fixed gun to be a liability. A Canadian prototype was built in twenty months, and the tank had a cast steel hull, a large revolving turret on a seventy-two-inch ring, and a 75mm calibre gun. It used an American engine. Not only did American defence officials agree that it was a sound model, but they also installed the Canadian turret and gun on what came to be the *M-4 Sherman* tank.[39] But as with the *Lancaster* and the *Tribal* destroyers, Canadian inexperience resulted in lengthy production times. There were also engineering and armour plate problems as well as a high final cost for the *Ram* relative to the *Sherman*. The latter subsequently became the main battle tank of the Allies, and the Canadian armoured divisions were equipped with *Ram* tanks only until they could be replaced by the American model.[40] Although it was correct to switch to the *Sherman* due to the inability of domestic industry to produce an army tank quickly and cheaply, the disadvantages of the purchase over the production of military equipment became obvious after the war. In 1949, Minister of National Defence Brooke Claxton was forced to write to the US Secretary of Defence as a last resort to try to procure improved bogie wheels and treads for Canadian *Sherman* tanks. The attempt failed because it was claimed that they were needed for US requirements.[41]

Canada did have the option to purchase some military materiel in the United States during the war. In August 1940, Prime Minister King and President Roosevelt signed the Ogdensburg Agreement, which established the Permanent Joint Board of Defence to facilitate discussions on the defence of the continent. The Hyde Park declaration was subsequently made in December 1941. It consisted of the approval of a statement by the Joint War Production Committee of Canada and the United States that called for a combined production effort.[42] The relationship with the United States regarding joint defence production during the Second World War, however, did not completely replace that with the British. J.L. Granatstein is correct in his observation that

> The events of the war had marked a historic change in Canada's place in the world. Curiously, however, there was as yet little sign of this in the attitudes, equipment, and training of the Canadian forces. The direct influence of the American military on their Canadian counterparts was still relatively limited. The methods and models for Canadian soldiers, sailors and airmen without question remained British in 1945.[43]

This trend gradually changed in the postwar period. The Canadian government had made the decision to standardize on American patterns of military materiel in 1947, and all three Canadian military services subsequently moved, to varying degrees, toward that goal.[44] As Peter Archambault has explained, however, the defence ties with Britain remained substantial despite their lack of formality, and Canada still procured British equipment – the aircraft carrier *Bonaventure* and the *Centurion* tank – when it was to its advantage.[45]

By the Second World War, many defence products were built in Canada and in tandem with its new continental partner. Although the Canadian government did finally accept the concept of military preparedness in peacetime to a small degree after the war, the Canadian military was still forced to rely on a combination of foreign options due to a failure to maintain its defence industry.[46] Although the transfer to the American production scheme was due to proximity and burgeoning economic co-operation between the two nations, it was also a symbol of casting off the yoke of the British Empire and conducting business under more favourable conditions.

The Canadian government's endorsement in the late 1940s of the construction of an all-weather jet fighter interceptor – the *CF-100* – is one example of a Canadian effort to advance its defence industry and procure domestically. But the aircraft failed to make an impact on the international market because of a lack of Canadian industrial experience in defence production and meeting deadlines. The design proved that Canadian industry could create superior weapons technology, but its inexperience in manufacturing hurt the project. Canadian industry had also engaged in an overly ambitious project; one company tried to make its first-ever jet engine at the same time as a modern fighter. The company that was to build the new engines and the *CF-100* was A.V. Roe (Avro) Canada, and it began operations in November 1945. It was announced in January 1947 that the TR5 engine, later known as the Orenda, would be ready at the end of August 1949, and the *CF-100* could "tentatively" fly by that time.[47] Although this was a year later than first expected, it was progress never before seen in Canada. Colonel (retired) Randall Wakelam has written that, "indeed, whether the CF100 flew in 1948 or 1949, the fact that there were deadlines and aircraft at all was marked tribute to the foresight and determination of the air force."[48] Although Wakelam is correct that the innovation in aircraft was impressive, firm production schedules are vital to any defence acquisition.

Shortages of the Orenda engine plagued the possibility of deliveries through the first half of 1951. In fact, almost every aspect of the Avro operation was having extreme difficulties. Even after the first ten pre-production aircraft were delivered, a wingspan deficiency forced further delays. The main problem was that, although it had a good design team, the company lacked the expertise and experience to effectively turn the design into production capability and deliver on schedule.[49] The United States was originally interested in the *CF-100* as it was determined to be a high-quality fighter, but American officials were firm that they could only use the fighter if it was available before September 1952. The United States never acquired the aircraft. The first delivery of the *CF-100* was completed on 30 September 1953. By the end of the Korean War, only ninety fighters had been built compared with the 1,000 *F-86* fighters produced in Canada during the same time.[50] The Canadian government had spent almost $750 million from start to finish, and the only sale made of the *CF-100* was for fifty-three aircraft to Belgium in 1957.[51] Despite the increased budget allotments created by the Korean crisis, the Canadian defence industry was simply not ready to undertake such a project, and the aircraft were not ready when they were needed in 1950.[52] Wakelam admitted this later in his study: "There was no doubt that the CF100 was a world class aircraft ... The CF100 was the aircraft that the RCAF wanted to succeed. Sadly, AVRO's production capability proved to be too immature to be fully responsive to the changes in priorities and the crash programmes sparked by the events of 1950."[53] The necessary challenges that needed to be overcome to succeed in the defence industry were later explained in a Defence Research Analysis Establishment report: "The ability to achieve or retain an internationally competitive position is determined not only by the price at which the existing products are sold but also by the speed with which new and superior products and processes are introduced to the marketplace."[54] The *CF-100* failed to meet these criteria in the 1940s. The long history to that point of Canadian reliance on foreign sources made the defence industry largely unable to succeed in the independent production of advanced military technology.

A major exception to the reliance on foreign sources for equipment was the Canadian-designed *St. Laurent* class ASW escorts of the RCN. Radically new ships were needed to counter the threat of Soviet submarine innovations.[55] As a result, Captain R. Baker from the Royal Corps of Naval Constructors proposed that a new design be constructed in Canada based on the British *Intermediate* class destroyer but with a continuous forecastle to allow for more space for operations rooms and communication centres necessary for a modern vessel. The problem was that Canadian shipyards had no experience in preparing a ship design; they had always relied on the British for such complex tasks. Moreover, the technical offices at Naval Service Headquarters were not staffed as a warship-design authority. The Canadian government accepted the idea for a Canadian-designed ship and subsequently authorized the growth of the engineer-in-chief's department. The department expanded from five officers in 1948 to twenty-one, supported by civilian engineers, technologists, and project managers. A Naval Central Drawing Office was also created for the first time, and the Naval Central Procurement Agency was organized as an offshoot.[56] Davis concluded that "this new professional core gave the RCN the ability to design its own warships from scratch, rather than merely copying those of other Navies."[57]

The result of the new Canadian initiative was the *St. Laurent* class ASW escorts. The ships were based generally on a British design but completely redrawn for modern Canadian needs. In addition to the extra space for operations and communications, the design provided passageways and excellent access routes to facilitate the rapid closing down of the vessel. The *St. Laurent* ships were the first NATO ships to provide such arrangements for closure as a means of nuclear, biological, and chemical defence (NBCD). The design allowed the ship to continue fighting during a nuclear attack.[58] For the first time, the RCN had designed a major war vessel. The inexperience in weapons and equipment design was still apparent, however, as financial planning for the original program proved highly inaccurate and the final cost was at least three times more than that first projected. The naval planners also desired that domestic industry be able to fully support the ships, and this meant that the major systems had to be developed or manufactured in Canada. This requirement resulted in a myriad of delays as many of the pieces had to start on the drawing board, and indecision prevailed as to the final shape, weapons, and electronic suite to be used. The *St. Laurent* escorts were not operational until 1955 – three years behind schedule.[59] Such a situation places any procurement at risk of cancellation. But as Michael Hennessy has written, "the perceived likelihood of general war, however, distracted the government's attention from matters of ultimate cost."[60] The project was also marred by the fact that the ships were quickly shown to be ineffective against modern submarines; they were too slow, and the range of their hull-mounted sonar and anti-submarine weapons was too short.[61] They needed upgrades soon after their introduction to the RCN. Although the *St. Laurent* is a rare example of industrial independence in Canadian procurement history, it was apparent that having ships designed strictly for Canadian needs was a costly and time consuming venture.

The impact of American weapons systems in Canada after the Second World War was not immediate or total. When it came time to send the Canadian Army Special Force to Korea in the spring of 1951, it was equipped with Second World War British-pattern equipment.[62] The infantry's rifle was the bolt-action Lee Enfield .303 No. 4, Mk. I. The army's anti-tank guns, mortars, small arms, tanks, field artillery, radios, signals equipment, and helmets were all from the previous war.[63] Once in Korea, the Canadian army began to adopt American-pattern weapons and equipment in piecemeal fashion.[64] Although doing so placed Canada within a more recognizable North American framework and gave the Canadian government more freedom regarding its weaponry, it continued the trend of trying to adapt foreign designs to Canadian needs. Throughout the Korean War, the Canadian government began to reinvest in a domestic defence industry. One of the primary areas of investment was aircraft to meet the Soviet bomber threat. The desire for a domestic source for aircraft continued after the war, and the result was the penultimate procurement failure in Canadian history – the *CF-105* or, as it is more commonly known, the Avro *Arrow*.[65]

Even Canadian citizens with no interest in military history have heard of the Avro *Arrow*. And since it is perhaps the one topic in the procurement field that has received adequate academic attention, it will suffice to simply outline the events here.[66] By 1953, the *CF-100* was outdated and was no match for Soviet aircraft.[67] The answer was the *CF-105* supersonic, twin-engined, all-weather jet fighter. The project had been in development in 1949 and was accepted by the Liberal government in 1953. By 1955, the necessary redesigns of the airframe and fire-control system had put it vastly over budget. The Liberal government continued to pour money into the venture in the hope of creating the world's best jet interceptor to counter the appearance of the Soviet intercontinental bomber in 1954. But production delays continued. Worse still was the fact that nobody wanted to buy it; it was completely unproven, especially the avionics and weapons systems, and Canada still did not have a reputation for building jet aircraft. As with the *CF-100,* the lack of experience in designing and producing military materiel was a major liability. As Dan Middlemiss has correctly written,

> Thus by 1956, the development of the "Arrow" had burgeoned into an all Canadian programme despite the initial intentions by the government to keep Canadian participation strictly limited. As later events were to demonstrate, an attempt of this sort to develop and produce all the major components for a major weapons system by inexperienced manufacturers was almost predestined to fail ... No allowance was made for the inevitable development problems and delays that would be encountered in such an ambitious project.[68]

Former Chief of the General Staff Lieutenant General Guy Simonds criticized the project at the time for consuming too much of the defence budget and ignoring the trend toward ground-to-air missiles. He realized, however, that the desire of the air force, the aircraft industry, and defence research scientists to participate in a project that they could call their own swept aside any opposition to the venture.[69]

By 1957, the Liberals happily passed the problem to the new Conservative government under John Diefenbaker. On the same day that the first *Arrow* prototype rolled off the Avro line, the Soviets launched *Sputnik I* into orbit. The age of intercontinental ballistic missiles had begun, and the rationalization behind the *Arrow* quickly began to fade. On 20 February 1959, the Diefenbaker government scuttled the entire project, and Avro was forced to fire 14,000 employees instantly. The existing prototypes were subsequently destroyed without explanation. The initial reasons for the cancellation of the *Arrow* was that it was obsolete in the missile age and that the American Bomarc defence missile was more appropriate. The truth was that the aircraft had simply taken too long to produce and had become too expensive. The Diefenbaker government had little information on the project and less inkling to search for it. George R. Pearkes, the Minister of Defence at the time of cancellation, later gave this interpretation:

> We were defenceless against the high powered bomber – we had the old CF-100, it couldn't compete with the modern Russian bomber; we had no supersonic fighter, but the Americans emphasized the fact that they had lots

> of them. Now then, the question I had to face ... was, if you scrapped the Arrow, you'd get nothing; what will you do? Will you buy American aircraft to fill in this gap – cheaper American aircraft ... That's where people began to tear their hair and say "you scrapped the Arrow, now you're turning round and buying cheaper (and they would say 'not so good') aircraft."[70]

As Desmond Morton has asserted, "in power, he [Diefenbaker] had taken one hard look at the costs of technological independence, quailed, and fled."[71]

The *Arrow* was not the only attempt to design and produce a weapon system for export in the 1950s. Between 1948 and 1955, the Canadian Army Research and Development Establishment (CARDE) had developed a 3.2-inch medium anti-tank round called the *Heller*. It was considered by CARDE to be more accurate than the 17- and 25-pound "pot sabot" round and could be carried and fired by an individual soldier. The Canadian army adopted the weapon between 1956 and 1960. The real goal, however, was to have it procured by both the British and the Americans to improve the level of standardization within the North Atlantic Triangle. Canadian hopes were dashed when the British purchased the Karl Gustav from Sweden instead. The British felt no compulsion to support Canadian weapons development and did not defend the decision to Canadian officials.[72]

There was also the Canadian *Bobcat* Armoured Personnel Carrier (APC) meant to carry troops quickly into the field in the case of war. The design began in 1953, and by 1956 a prototype had been created. Although the Cabinet Defence Committee had approved its development that year, Canada was still unable to generate any interest from Britain or the United States by 1958. The Canadian army had hoped that it would be the standard NATO APC. The British and the Americans had their own models under development, however, and their projects had taken half the time to get to the same point. It was not until 1964 that the Cabinet Defence Committee finally gave up on the *Bobcat* project due to insuperable costs and remaining technical problems. The Canadian army then looked to the American model of *M-113*s for its APC requirement. As Peter Archambault has written, "the Bobcat and Arrow projects had met the same fate, for the same reasons: cost overruns; no foreign market; and the failure of the manufacturers concerned to make a successful transition from development to production of the prototypes."[73]

It was the failure of the *Arrow* program that had the most profound effect on the political will to design and produce weapons and equipment in Canada. After the *Arrow* program was first curtailed in September 1958, the Canadian government began discussions with the United States regarding the future of the Canadian defence industry; the conclusion was to forgo independent Canadian weapons production and enter into production-sharing agreements with American defence firms. Canadian military officials held minimal influence in this decision, and it was clear that the politicians were going to handle the procurement strategy. The government's objective was to increase the participation of Canadian industry in the production and support of weapons and equipment used in North American defence. But Canada could not afford to do so without some form of sharing agreement between the two nations. The Canadian defence industry was weakened by the *Arrow*'s situation and needed help from its neighbour. The American government agreed that this was an effective way for Canada to contribute to continental defence, and a series of Defence Production and Development Sharing Arrangements was concluded near the end of 1958.[74]

This is only a snapshot of the history of procurement in Canada up to the 1960s. It does, however, explain that Canada had, to that point, relied mostly on foreign sources for its military equipment. Canadian industry had rarely designed its own weapons. The most common course had been for the DND and the government to purchase foreign prototypes and then attempt to make alterations to them to meet Canadian needs. This practice had invariably led to delays in the procurement process. The attempt to create a domestic industrial base to tailor weapons and equipment to specific Canadian military requirements had also rarely met with success. Inexperience with weapons design and production had usually resulted in recurring delays and escalating costs. As with the case of the *Arrow,* the longer a defence project goes on in Canada, the more politically vulnerable it becomes. It was within this environment that the DND attempted to procure the *Sea King* helicopter.

Sikorsky *S-55*, or H04S *Horse* Helicopter.

Source: DND photo. *Courtesy of National Defence. Reproduced with the permission of the Minister of Public Works and Government Services, 2009*

2
Early Helicopter Operations: The Exploration of a New Capability

The true beginnings of naval aviation in Canada came from the Second World War, when Canada became the third largest surface navy in the world. The Royal Canadian Navy (RCN) had recruited air crew for the Royal Naval Air Service (RNAS) in April 1915, and more than 600 joined to fight against the German U-boats. The RNAS was disbanded, however, when it was amalgamated with the Royal Flying Corps to create the Royal Air Force (RAF) in 1918. In September 1918, Canada created its own branch called the Royal Canadian Naval Air Service (RCNAS) to guard its East Coast. But with the signing of the Armistice in November 1918, the new RCNAS was "discontinued," and the concept was abandoned due to budget cuts in the interwar years.[1] Even during the Second World War, Canada did not have a naval air arm or aircraft carrier; in fact, the first planes used by the RCN were four different types of biplanes. By 1943, Canadian pilots had begun to train and serve with the Royal Navy (RN) Fleet Air Arm.[2] Near the end of the war in 1945, they were preparing for action in the Pacific with the 19th Carrier Air Group within the RN. And notwithstanding the end of the war on 10 August, three of the four Canadian squadrons that would form the postwar core of Canadian naval aviation were in existence by the end of the month.[3]

Plans continued toward the formation of the postwar shore base for naval aviation in Canada, and by December 1945 the Royal Canadian Naval Air Service had been formally created at the Royal Canadian Air Force (RCAF) base in Dartmouth, Nova Scotia. On 19 December, cabinet approved, in principle, the formation of an air component within the RCN, and two days later final authority was given for its establishment. Perhaps most importantly for the new air arm was that the aircraft carrier Her Majesty's Canadian Ship (HMCS) *Warrior* was subsequently commissioned on 2 January 1946. On 31 March, it entered Halifax Harbour and made the Royal Canadian Naval Air Facility at Dartmouth operational. Although this was a clear commitment to establish a credible postwar RCN, the history of weapons procurement in

Canada remained consistent as the navy had taken possession of a ship that had not been built for its requirements.

HMCS *Warrior* had been built to operate in the warm weather of the Mediterranean and the Pacific. The main area of responsibility of the RCN, however, was the North Atlantic, and the *Warrior* did not have adequately winterized living spaces for these conditions. The carrier was subsequently sent to the West Coast via Panama to enter the Pacific and friendlier temperatures near Esquimalt. This was the last winter that the *Warrior* was in service in Canada. On 14 January 1947, the Canadian cabinet decided to return the carrier to the British and purchase one more suitable to Canadian needs. This decision ended the hope of the RCN of operating two carriers. Although it was cited that the return of the *Warrior* was due to its inability to operate effectively in the North Atlantic, one historian has claimed that it was purely a political decision based on cuts to the defence budget.[4] Indeed, the RCN was being pared down from its wartime form. The White Paper entitled *Canada's Defence 1947* focused on demobilization, and Prime Minister Mackenzie King was determined to focus on social programs over defence.[5] As a result, the hopes of naval planners for sustained expansion and a task force that centred on light fleet carriers were weakening. By March 1946, the RCN had decommissioned 346 ships. As W.G.D. Lund has written, "in essence, the RCN temporarily ceased to function as an operationally capable fleet. The bold plan enunciated in 'The Continuing Royal Canadian Navy' grounded on the hard reality of the immediate postwar climate where neither strategic requirements nor alliance commitments existed to justify a substantial navy."[6] One carrier was to be maintained, however, and in April 1948 the *Warrior*'s replacement, HMCS *Magnificent,* was commissioned and set sail for its new home in Canada. It would be the floating base for British-designed *Sea Fury* and *Firefly FR IV* aircraft. By that time, the moniker "Naval Air Arm" had been discontinued, and "Naval Aviation" had taken its place. The base at Dartmouth was also transferred from the RCAF to the RCN in 1948 and made the navy "master of its own house" in the air. The base was commissioned Royal Canadian Naval Air Station, Shearwater, and became the headquarters of Canadian naval aviation.[7]

The North Atlantic Treaty was signed on 4 April 1949. Prior to this event, American foreign policy was heavily influenced by Canadian diplomatic pressure concerning the creation of a multilateral security agreement. The year before the pact was signed, Lester Pearson, the Canadian Secretary of State for External Affairs, stated that "the North Atlantic Community is today a real commonwealth of nations which share the same democratic and cultural traditions. If a movement towards its political and economic unification can be started this year, no one can forecast the extent of the unity which may exist five, ten or fifteen years from now."[8] The period surrounding the NATO discussions was characterized by rapid technological change and is often referred to as the "postwar naval revolution." The strategic environment then included the fast submarine, guided missiles, and, of course, nuclear weapons. The coup in Czechoslovakia in February 1948 and the blockade of Berlin in March had increased the likelihood for defence planners that Soviet submarines might soon be used for an attack against shipping. Although the Soviet surface fleet was relatively weak, Western defence planners believed that the Russians had mastered the advanced submarine technologies captured from Nazi Germany. By the late 1940s, the Soviet submarine fleet was the largest in the world, and both the United States Navy (USN) and the Royal Navy were restructuring their forces in response to that threat. They focused primarily on offensive capabilities against shipping and choke points.[9] The Canadian government, however, determined that the RCN's role within NATO would be to expand its Second World War tasks and specialize in Anti-Submarine Warfare (ASW) and convoy protection.[10] Members of the Canadian Naval Staff had actually been discussing much earlier how the primary commitment of the RCN was ASW.[11] Commander A.H.G. Storrs, Assistant Director of Naval Plans and Intelligence, had written a paper in January 1947 calling for the service to concentrate on the development of fast escorts to counter submarines in response to the imposition of a manning ceiling and continued budget cuts.[12] As Lund has written, an ASW focus frustrated the postwar plans of those within the RCN leadership who were arguing for carrier-oriented general purpose task groups.[13] The political will to invest in well-balanced Fleet Task Forces during peacetime, however, did not exist. The memory of the toll exacted by enemy submarines during the war, however, was still fresh, and a continued defence against it was more easily justified. The official ASW direction, therefore, allowed RCN officers to maintain and expand these Second World War capabilities.[14] Brian Cuthbertson has argued that the rebirth of the postwar RCN was dependent on the creation of NATO.[15] The problem was that, although the commitment to NATO provided the impetus for naval expansion, the RCN quickly overcommitted itself regarding the number of anti-submarine escorts to be used within the alliance.[16] The commissioned fleet in early 1949 consisted of one light fleet carrier, one cruiser, five destroyers, two frigates, and one minesweeper.[17] Soviet submarine experimentation and development had also created a fear in the West regarding the obsolescence of Western surface vessels. There was a realization that new measures in the field of ASW had to be created as they had not kept pace with advancements in Soviet submarine technology.[18] Although the RCN had a definite role after the signing of the NATO agreement, the means of its execution were less clear.

Prior to 1951, the RCN's naval aviation arm was composed of strictly fixed-wing aircraft.[19] But a new platform was about to have a major impact on how the RCN carried out its Search and Rescue (SAR), aerial communications, and ASW roles – that of the helicopter. The very idea of operating helicopters from Destroyer Escorts (DEs) in Canada dates back to a memorandum from 1943 by Lieutenant D.W. Overend, Staff Officer (Fuel), recommending that "some of the Canadian frigates under construction should be completed as anti-submarine helicopter carriers."[20] The RN, however, largely believed that helicopter operations should only be based onboard aircraft carriers. British Lieutenant Alan Bristow had landed a Sikorsky *R-4* aboard Her Majesty's Ship (HMS) *Helmsdale,* a *River* class frigate, in September 1946. This landing, however, was more to impress some of the Royal Family in attendance, and helicopter trials were invariably carried out aboard British carriers.[21]

In January 1948, the Canadian Naval Staff approved the requirement for an icebreaker to carry helicopters during its operations in the Arctic.[22] The icebreaker itself had not been procured yet; however, once the Naval Board agreed to accept helicopters over fixed-wing aircraft, this decision gave impetus to their acquisition.[23] The icebreaker, therefore, was a major reason the navy began to procure helicopters.[24] The contract for the icebreaker was approved in January 1949. Although it would be an American design, what became HMCS *Labrador* was built in Canada at Marine Industries Limited, in Sorel, Quebec, and was modified to accommodate up to three helicopters. The project was delayed as the Canadian manufacturer waited for the delivery of material and equipment for assembly of the ship. Canadian alterations to the US design exacerbated this delay and deferred completion of the ship until December 1951. As a corollary, the procurement of helicopters was postponed. Notwithstanding the delays, it is clear that the RCN was progressive regarding the possibilities of rotary-wing aircraft, as even the RCAF had only procured its first helicopter in 1948, the Sikorsky *S-51,* later renamed the *H-5,* used primarily for SAR.[25]

The Canadian Commander, Canadian Destroyers Far East, A.B. Fraser-Harris, also contributed to the concept of DE-based helicopters while aboard HMCS *Nootka* during the Korean War. He recommended to Naval Service Headquarters in 1951 that all warships larger than corvettes be equipped with flight decks. Captain W.M. Landymore, Commanding Officer (CO) of HMCS *Iroquois,* made a similar recommendation the following year:

> This, of course, is not a new thought or procedure to either the RCN or RN but is felt that such great strides forward have been made in their operations that it is time to enter the development with more than observers' interest ... *The most significant advance in A/S detection operations since World War II is this development and ... it holds the greatest possible promise for the future* [emphasis in original].[26]

Indeed, the first actual landing of a helicopter on a Canadian destroyer took place that year, when two RCAF Sikorsky *H-5* helicopters, one of which was carrying an injured soldier, landed on top of HMCS *Sioux*'s weapons handling room. The "observer interest" would remain dominant for a few more years, however, and the wedding of large helicopters to DEs was not initiated. It was not yet understood how it could be practically done. It is fascinating to note that the first Staff Requirements for naval helicopters – or Statement of Requirement (SOR), as it would later be called – were not even two pages long. Naval officials simply did not have any experience with helicopters and were unsure of what they could ask for. All that was written concerning ASW was "possible addition of an A/S weapon at a later date."[27] There was also a great deal of logical skepticism and hesitancy within the service in the beginning regarding helicopters. Their introduction into the navy was, therefore, very slow and required extensive trials and analyses.[28]

The first venture into the field of rotary-wing aircraft was with the creation of No. 1 Naval Helicopter Flight at Shearwater Naval Air Station in September 1951. It was composed of three American Bell *HTL-4* helicopters. One author has claimed that, by the end of that year, "the value of naval helicopters was wholeheartedly embraced by the Department of National Defence."[29] It is interesting to note that a Canadian prototype, the Sznycer *SG Mk. VI-D,* was originally considered over the Bell model. It was a single-rotor aircraft and was the first helicopter to be designed and built in Canada. It was first flown in 1947, but the firm went out of business shortly after the initial production.[30] Although helicopters were showing signs of utility at sea, the discussion on their place in naval warfare had only just begun.

The initial idea was to operate helicopters aboard a Canadian aircraft carrier. Cabinet had already approved the purchase of a British light fleet carrier on 23 April 1952. It was originally named HMS *Powerful* and was launched in November 1943 but had been docked in Belfast since May 1946. The decision was made, therefore, to procure an old and stagnant carrier and attempt to modify it to keep costs down. This was, after all, the Canadian way to acquire weapons and equipment for its military. The carrier was later named HMCS *Bonaventure* and was entirely Canadian owned.[31] Its purchase also followed the recurring theme in Canadian procurement history in that the military capability being acquired is secondary to the political reasons for it. As David Bercuson has asserted, "in Canada, political considerations have long played a key role in the procurement process. Her Majesty's Canadian Ship *Bonaventure,* the last Canadian aircraft carrier, was built in the United Kingdom and acquired by Canada in part to maintain Canada-UK defence ties at a time when Canada was starting to turn to the United States for many military requirements."[32]

The Canadian military was predisposed to procuring its helicopters from the United States. In 1952, the Canadian government approved the purchase of the American Sikorsky *S-55,* first acquired by the USN, with the designation *HO4S.* The first *HO4S,* or *Horse* as it came to be known, arrived in Shearwater early in 1952 and was initially used to fight forest fires in the Atlantic provinces.[33] While the *Horse* began contributing to fire control, it made its first landing aboard the *Magnificent* on 6 May 1952 in order to function as a "plane guard." This was a rescue mission to retrieve the air crew from any aircraft that ditched over the side of the carrier. This mission was normally performed by destroyers. In December of that year, No. 1 Naval Helicopter Flight was elevated to squadron status and designated VH 21.[34] It operated three Bell *HTL-4*s in a utility role and three *HO4S-2*s in a general purpose and plane guard role. The RCN had also purchased one navalized version called the *HO4S-3.*[35] The *HO4S* models quickly excelled in their tasks, which continued to be expanded within the Canadian navy. J.A. Foster recalled that

> The first RCN helicopter rescue at sea took place off Newfoundland on a cold October day in 1953 during NATO exercise "Mariner" when Lt. David H. Tate signaled HMS *Magnificent* that his Sea Fury was losing power. He ditched during his approach a half mile short of the flight deck. The airborne Sikorsky on "plane guard" reached him in 32 seconds and lowered its rescue cable. Moments later a cold and dripping Tate was back on board the "Maggie" warming himself in the wardroom.[36]

Clearly, there was much potential for helicopters within RCN operations.

The Canadian Naval Board met and decided on recommendations made by the Naval Staff. The board then passed its recommendations to the chiefs of staff, who then passed on their conclusions to the Minister of National Defence through the Deputy Minister. The core of the Naval Board consisted of the Chief of the Naval Staff, the Vice Chief, the Navy Comptroller, the Director of Personnel, the Assistant Chief of Naval Staff (Plans), and the Assistant Chief of Naval Staff (Air and Warfare).[37] In a special meeting called on 16 November 1953, the Naval Board discussed the topic of Canada's ASW role, including an examination of whether helicopters would contribute to "seaward defence."[38] A committee was subsequently established in April 1953 to investigate the requirements for helicopters, and as a result of its conclusions the Naval Staff gave approval on 6 October for the formation of a squadron to be used in ASW. Two general advantages of helicopters were described as a reduction in time for investigating contacts at sea and an augmentation of the surface screen.[39] The Chief of the Naval Staff stated that the use of helicopters could be "a marked step toward making up the deficiencies in the lack of surface escorts."[40] This conclusion centred on the new developments in dipping sonar technology – specifically the AN/AQS-4 – that the USN proved could locate and track submarines.[41] The AQS-4 involved lowering a sonar transducer on a cable to sixty feet below the water's surface as the helicopter hovered ten to twenty feet above the water. If a submarine was in the vicinity, the sound waves generated would give a return echo and give the position of the sub. The range of the sonar was 1.6 times that of the surface ships. The Naval Board then gave their approval of a helicopter squadron composed of the *H-21* model by Piasecki. This American model was a large dual-rotor aircraft considered ideal for the RCN because it could carry both sonar and the Mark 43 torpedo. The Canadian chiefs of staff, however, refused the Naval Board's recommendation, agreeing only to allow the RCN to form a small provisional unit for experimental purposes. They stated that it could be formed when more funds were made available from the naval estimates.[42] A lack of money was part of the reason for this decision. But hesitancy also remained within the RCN in regard to helicopters, and more investigation was necessary before any firm organizational commitments could be made.

At the October 1953 meeting of the Naval Board, Vice Admiral Geoffrey Barnard, who represented the British Admiralty, said that they too were particularly interested in the new role of ASW helicopters and that perhaps the day would come when the helicopter, "suitably equipped with dunking Sonar or newer equipment, would take over entirely the role now performed by the A/S escort."[43] He believed that the concept only needed a more dependable helicopter standard that could fly under all weather conditions. The icing of the blades in cold weather operations was of particular concern. Barnard mentioned that the United Kingdom was behind in helicopter production, however, as it had been concentrating on building other aircraft and had not had the time to conduct evaluation trials in this developing field. At the meeting, the question was posed to Barnard: "Is operation of helicopters from small ships now ruled out and if so why?" He stated simply that it was unlikely that helicopters would fly from smaller escort vessels. His first statement was revealing, as Canada did indeed not look to Britain for its purchase of helicopters but to the United States. His second comment about aircraft flying from smaller ships was also enlightening, as it showed that there was a myopic view within the RN about the possibilities concerning rotary-wing ASW warfare in the early phases. At the meeting that November, however, the Canadian Assistant Chief of the Naval Staff (ACNS) revealed that Barnard's view was not universally held in Canada. He stated that no decision had yet been made as to whether helicopters would be used under war conditions from suitably fitted merchant ships or from operational carriers. Notwithstanding where they would operate from, the

ACNS concluded in 1953 that it was hoped the RCN would shortly acquire an operational squadron of six aircraft.

By December 1954, a contract demand was let for the purchase of six new Sikorsky *HO4S-3* ASW models.[44] As the helicopter component grew in actual size and, concomitantly, the importance of research and development on ASW, the Naval Board debated the possibility of acquiring a second aircraft carrier. There were many reasons for this possible acquisition. The Vice Chief of the Naval Staff (VCNS) tabled a paper on 19 May 1954 stating that "the RCN commitment to NATO to provide one carrier could not be met with one carrier only, as she would be non-operational approximately 50 percent of her time, due to refits, aircrew training and exchanging of squadrons."[45] It also concluded that one carrier provided the RCN with a proportionately much smaller air arm than that of any other navy with an air component and that, if the one carrier was damaged, the entire RCN air arm would be non-operational. It was also clear to the Naval Board that the USN or the RN would not be able or willing to provide Canada with an operational carrier in the event of a crisis since they were already in short supply themselves. The board thought that, if HMCS *Magnificent* were retained, in addition to the commissioning of HMCS *Bonaventure,* then *Magnificent* could be used as a "helicopter mother ship." These conclusions, of course, were based on the belief that helicopters could fly only from carriers, and the board approved a further study into acquiring a second carrier for the fleet.

By July 1955, the first ASW squadron had been formed. It was the small ASW Helicopter Test and Development Unit that the chiefs of staff had approved in 1953. It was called HS50, and the unit was based in Shearwater. It was outfitted with six dipping sonar-equipped Sikorsky *HO4S-3* helicopters. The new helicopters facilitated the development of ASW tactics and procedures aboard the light fleet carriers HMCS *Magnificent* and its successor, HMCS *Bonaventure.*[46] Once the ASW squadron was formed, the Sikorsky *HO4S-3* became the foundation of naval helicopter operations in Canada.

Exercises conducted from October until the end of the year demonstrated that ASW helicopters had a major role to play in seaward defence. These exercises included screening inbound and outbound convoys in the harbour approaches and channels with the help of the onboard AN/AQS-4 sonar. As Michael Shawn Cafferky has stated, "wartime experiences had shown that submarines could lurk in the complex waters off Halifax, and shore-based helicopters, because of their ability to hover and dip sonar, were much better suited than fixed-wing aircraft to the precision work of tracking submarines close inshore."[47]

Discussions concerning the procurement of a more modern helicopter had begun the previous year, in 1954, and a draft for Staff Requirements was presented to the Naval Staff. It is fascinating to note that there were no helicopters in existence that fulfilled the proposed requirements, but the staff concluded that they were still "technically realistic."[48] This demonstrated refreshing forethought within the RCN with regard to the role that the helicopter could play in the future of naval aviation. The minutes of the meeting revealed this further: "Therefore, ACNS (A) proposed that the Staff Requirements should now be approved in order that further consideration could be given to exploring the possibilities of eventual manufacture in Canada of a suitable ASW helicopter for the RCN if and when such acquisition was approved." Although this statement does not mention anything about actually designing the aircraft, the facts that the staff approved a Staff Requirement knowing that the piece of equipment did not even exist, and considered the possibility of making it in Canada once the technology was available, are exactly how staff work should be done – looking to the future and what role the nation's military will play within it. There was even discussion on whether there would be two types of helicopter, one for SAR and the other for ASW. The ACNS concluded, in hindsight correctly, that a "dual function machine would make the employment of helicopters more flexible and was likely to enable them to be used in tactical situations not at present visualized."

The Staff Requirements document for a new helicopter, first drafted in 1954, continued to evolve as a result of the changing strategic environment. The NATO strategic plan during the 1950s, titled MC-48, predicted a war of short duration that would involve a struggle to gain atomic superiority.[49] Intelligence estimates concluded that, by 1960, Soviet long-range submarines could bring strength to bear on the North American coast and threaten the US atomic striking force, a large portion of which was within 100 miles of the sea.[50] In 1954, the Western strategy to deal with the Soviet submarine threat involved the construction of Sound Surveillance Systems (SOSUS) at sea. SOSUS was a series of acoustic receivers called "arrays" embedded in the ocean floor in a chain over several hundred miles. They were meant to monitor when a Soviet submarine entered the wider Atlantic between either Iceland and the United Kingdom or Iceland and Greenland. It was a joint project of the USN and Bell Laboratories, and Canada joined the project shortly after its inception in late 1954.[51] As Cafferky has written, "Strategic Surveillance became the watchword for the RCN."[52] Although SOSUS was theoretically an effective system to secure the approaches to North America, a modern fleet was required to respond to a successful enemy contact. The RCN ASW forces were required to patrol an area approximately 700 by 300 miles. Twenty-four ships were needed to operate in pairs to cover the search area. The new class of *Restigouche* ships was expected to be commissioned by the end of 1958. The RCN, therefore, expected to have fourteen *St. Laurent/*

Restigouche DEs; and four *Tribal* class destroyers in service by that time. There were also the original four British-built *Tribal* class destroyers and seven *Prestonians,* but none of these possessed the speed or adequate ASW gear for SOSUS contact missions. This meant that the RCN was six ships short of reaching the required twenty-four long-range patrol vessels. Ship procurement, therefore, became a priority.[53] Although it was recognized by some naval officials that helicopters would play a major role in this new context, they were still unproven in Canadian ASW, and it was believed that they could operate only from a carrier. The RCAF also questioned the use of helicopters. Naval Staff were skeptical of the navy's future assessments because they believed that fixed-wing aircraft, in conjunction with SOSUS, could reliably detect Soviet submarines.[54] These factors continued to sideline any approval of funding for a major helicopter project.

Notwithstanding the model to be chosen or what year the money was to be found, it was clear that the new ASW helicopter would not be designed in Canada. This was not because Canada lacked the confidence to create major domestic equipment; the Avro *Arrow* failure had not yet occurred. And even though the *St. Laurent* class had been behind schedule and over budget, the RCN was still very proud of the Canadian-built ships.[55] Although optimism for building major Canadian equipment platforms existed, it simply did not transfer into the field of helicopters, as the larger navies were pushing ahead much faster. The reason stated by the Chiefs of Staff Committee in 1956 for the rejection of helicopter manufacture in Canada was that only a small number were required; it would not make financial sense, it was believed, to invest in design and production for six aircraft. Perhaps more importantly, Canada discarded its own design and production concept because a new American model that could satisfy the developing Canadian requirement emerged – the Sikorsky *S-58.* Although the chiefs of staff were correct to pursue an existing helicopter rather than attempt to design a Canadian version, they were limited in their understanding of the role that the helicopter would play and how many would be needed to facilitate the new role of Canadian ASW within NATO. Another reason for the delay in selecting a helicopter and approving funds was that the chiefs of staff were awaiting the conclusions of a DND committee exploring whether they could coordinate future helicopter requirements for all three services and create joint training. Although it was concluded in March 1956 that training would be separate and that RCN pilots would continue to train at Shearwater, it was obvious that there was a need for more helicopters for the Canadian military. The decision to keep them separate was due to the necessity of unique features for a naval helicopter, such as taking off and landing on a floating ship, whereas the army required a helicopter to carry heavy and bulky loads.[56]

One reason the Naval Board believed that the *HO4S-3* helicopter had to be replaced was that the experimentation squadron, HS50, was not providing significant results.[57] The RCN was committed to being a global leader in ASW warfare and did not want its technology to limit its aspirations of innovation in the field. Nowhere was this aspiration more apparent than in the endeavour to expand the ASW helicopter/ship platform and fly helicopters from independent escorts and destroyers. Helicopters were very limited within the RCN if they could operate from only one base at sea. Although helicopters had proven that they could be competent in an ASW role, the Chief of the Naval Staff (CNS) had stated in February 1956 that "an ASW Squadron of 12 would require a special Helicopter ship to operate from. No authority exists to provide such a vessel."[58] Canadian maritime warfare officials, therefore, began to consider what they had to work with and the possibility of placing the aircraft aboard a smaller ship deck. After all, Canada had only one carrier, and, if the aircraft could be fit to fly from escorts and destroyers, this adaptation would vastly increase the capability of the RCN. It would effectively create what would later be termed a force multiplier. This idea was soon to inspire many within the RCN and was the beginning of a permanent breakthrough in international ASW operations.

On 23 January 1956, a Bell *HTL* helicopter from HS50 landed on the destroyer *St. Laurent*. It is unclear whether this was a test to determine the feasibility of wedding the two platforms into one, but credit for the idea of constructing a flight deck to operate helicopters from a destroyer must go in part to the ship's Executive Officer, Lieutenant Commander Pat Ryan, who was actually aboard the helicopter when it made the landing. Although it was thought that smaller helicopters might one day be able to operate from DEs, even the *HO4S* was considered much too big to fly from the back of a ship. These landings were, therefore, clearly anomalies within normal operations, and the concept of a large ASW helicopter-carrying DE was still completely embryonic.[59]

Although there was some discussion on disbanding HS50 prior to 1959-60, it was pointed out that, "although it has been ascertained that the RCN would gain from having helicopters associated with escorts, the question of how it can be done is still a problem and, therefore, there is still a requirement for HS50 to continue to be maintained at an operational strength of 6 helicopters."[60] There was still not sufficient data on what type of helicopter, if any, could operate in conjunction with an escort/destroyer vessel. The VCNS stated that, in his estimation, HS50, although termed an experimentation squadron, was in reality a nucleus unit with the long-term objective of training air personnel in the operation of helicopters. He also believed that it was to provide the ships in the Atlantic Command with ASW experience.

This was true, but the experimentation function of the unit was about to surprise its critics and become paramount in the solution to how helicopters could function together with escorts.

Although Canada was not leading the pack regarding helicopter design and production, HS50 was very active in the development of ASW tactics. At the beginning of 1956, Captain C.P. Nixon, Director of Tactics and Staff Duties, was concerned about the inability of existing DEs to counter modern submarines and stated that helicopters were "an essential partner in the ASW team; there is, in fact, no question that an escort with its own helicopter would be a far more efficient anti-submarine vessel."[61] It was decided to install an experimental helicopter platform aboard a DE, or "frigate," for the purpose of testing landings and take-offs; there was no requirement for maintenance or refuelling aboard the ships at the time. There had already been limited landings on HMCS *St. Laurent* in May, but its landing area consisted only of plywood fitted over the mortar well hatch. New trials were cleared to take place aboard the *Prestonian* class frigate HMCS *Buckingham* and began in September 1956. They included a steel landing platform welded to the quarter deck. Most of the trials were to ascertain the maximum degrees of pitch and roll that the helicopter could operate under for landings and take-offs, but they were also meant to measure the ship's ability to track the helicopter in poor weather and reduced visibility. As Peter Charlton, a former senior technical officer in Experimentation Squadron Ten (VX 10), and historian Michael Whitby have pointed out, "these were not meant to be operational ASW trials, but rather careful walks in a previously unexplored 'no man's land.'"[62]

VX 10 was established in March 1953 to develop and test military equipment to keep pace with emerging technology in airborne navigation and ASW for the RCN.[63] It included the Flight Test Division, the Project Engineering Division, and the Naval Detachment at the Winter Experimental Establishment at Namao. The 1956 trials were considered a great success, and they revealed that the helicopter-carrying destroyer (DDH) concept was feasible and that a brilliant force multiplier could effectively be created. This achievement was of paramount importance as, even with the *St. Laurent* class vessels entering service, the RCN still did not have enough ships for its NATO commitments, and naval officials were concerned about obsolescence. Although thirty-five escort vessels were due to leave the RCN between 1958 and 1968, the planned replacement program proposed in early 1956 would replace only twenty-six. And despite the high cost of the *St. Laurent* class that became operational in 1955, the ships were not highly effective against modern submarines, and refits were contemplated almost immediately after delivery. It was expected that even the newer *Restigouche* class, to be completed in 1957-58, would encounter the same weaknesses in its ASW capability.[64]

Helicopters were a clear cost-effective way to mitigate the lack of operational Canadian ships and could work in combination with them and fixed-wing aircraft to support SOSUS.[65] One issue, however, was the necessity of a hangar on the ship to protect the aircraft from saltwater spray to prevent long-term corrosion and allow a space for maintenance and repairs.[66] But far more importantly, the landing trials showed that "the limiting factors on helicopter operations were not in the pilots [sic] ability to land on a small moving platform, but rather on the ability to rapidly secure the helicopter to the deck."[67] The unavoidable movement of a ship immediately following the landing created a dangerous situation for those in and around the unsecured aircraft. Clearly, there was still much work to be done regarding the wedding of Canadian ASW helicopters to DEs.

The British had made attempts at a rapid-securing device for helicopters as early as 1948, including an idea to fit four interconnected "sucker pads" to a helicopter's undercarriage. They were to be powered by an engine-driven vacuum pump. The pilot would theoretically control the valve to release the helicopter from the flight deck.[68] There was also an idea in 1955 to have a harpoon grid system, where a barbed harpoon attached to the bottom of the aircraft would interlock itself to a grid suspended above the actual landing deck. But because both the USN and the RN were not seriously looking at flying large helicopters from smaller ships, these developments were temporarily suspended.

In the summer of 1954, H.B. Picken, the president and aeronautical engineer of the Genaire Company Limited of St. Catharines, Ontario, submitted a proposal to the Department of Defence Production (DDP) for landing a helicopter on a ship in rough conditions. It was a very complicated system that included a cable being lowered from the aircraft that was attached to a gyro and accelerometer in an attempt to keep the aircraft level. Although the system seemed to have potential, it was determined by the Defence Research Board during a meeting with Genaire that it was too intricate to develop. During the meeting, though, Deputy Director of Air Engineering Lieutenant Commander J.H. Johnson suggested that the communication link be removed and that perhaps a strong cable could be dropped from the helicopter and used to simply winch it in from the shipdeck; this was the beginning of a Canadian-designed haul-down and rapid-securing device.

The initial work of testing the reliability of attaching a cable to a helicopter and then using it to pull the aircraft down onto the ship was done by a group of engineers at Naval Headquarters under the direction of Commander John Frank. The testing of the concept was subsequently carried out by VX 10. Before any cable design could be initiated, VX 10 would have to prove that a helicopter could fly safely while being pulled by a tensioned wire. Bert Mead of VX 10 recalled that

> We had a Directive, which was rather an informal one, and I think that there was a bit of argument in Ottawa or somewhere about this being a stupid idea, because I recall I got a phone call in VX10 asking us to give some ideas on pulling a helicopter down. At any rate I went down to the old parade square down by C Hangar, we got a truck, rigged up a bunch of pulleys and all that sort of garbage and Bill Frayn came down, with an HO4S I believe it was, and we hooked on to him and the idea was to pull him down and see whether he lost control ... After a couple of times when Bill cut loose because he was losing control, we finally got it down to the point where he could sit there, steady as the devil, and we could pull him down ... So this really was the beginning of that whole concept of operations.[69]

The groundwork for a new type of naval platform had begun, and Canadian pilots and engineers were the pioneers. Notwithstanding the fact that the *HO4S-3* had proven worthwhile for testing it, its technology was beginning to lag behind that of the newer versions of ASW helicopters. It also had a high centre of gravity that caused it to pitch back and forth when landing; this was obviously a liability when trying to land within a very limited space. The undercarriage of the aircraft was also too weak to resist the pressures of the rolling and pitching of the deck. The RCN, therefore, continued the process to replace it with a more modern aircraft, and it was understood that it would be needed if trials were to continue.[70] The Staff Requirements set for 1960, first drafted in 1954, were indeed demanding and included new technologies, such as power-folding rotor blades, anti-icing, and an auto pilot and all-weather capability.[71] As the platform for combined arms at sea evolved, therefore, the navy decided that the helicopter chosen to operate from escort vessels should as well.

At the 546th meeting of the Naval Board on 24 July 1957, the naval comptroller stated that there was no money available in the 1957-58 estimates to initiate procurement of the six new helicopters under discussion.[72] As Cafferky has elucidated, "the trend towards the RCN's specialization in ASW was driven by political pressure, fiscal concerns, and Soviet advances on submarine and ballistic missile technology. Convincing the RCAF and the politicians of the merits of the helicopter to counter the Soviet threat was another matter."[73] By 1956, the cost of the Avro *Arrow* project was already under attack in the House of Commons, and Conservative defence critics challenged future military expenditures by the Liberal government. A recession had also begun in 1957, and it limited spending on the military in general for years to come.[74] Finally, a large part of this inability to secure money for major capital equipment purchases was due to the fact that the RCN received only 18 percent of the money spent on the three services between 1951 and

1960.[75] Under these circumstances, the navy could barely hold on to the concept of naval aviation and simply could not secure the necessary procurement from cabinet to take ASW to the next level by buying the *S-58*.

Notwithstanding the lack of a modern ASW helicopter, the experiments wedding helicopters and DEs continued. Previous exercises had focused on the carrying and landing of a helicopter on a DE. But in April 1957, tactical trials were conducted off Key West, Florida, to determine the benefits to be gained by employing ASW helicopters as part of a *St. Laurent/Restigouche* class vessel's weapons system. A summary of the report stated that, "from the limited trials conducted, considerable tactical advantage can be obtained from the use of helicopters in that, in suitable weather, they have the capacity to improve the range at which lethal weapons can be launched, or improve the rate at which doubtful sonar or radar contacts can be investigated."[76] In September, the Naval Staff continued to expound the impact of changing the face of helicopter-based ASW:

> The concept of embarking helicopters in escort vessels in the RCN was evolved to improve the ASW effectiveness of HMC Ships in locating and attacking the fast submarine ... The concept was also particularly attractive to the RCN in view of the difficulty of embarking helicopters in BONAVENTURE in addition to her complement of fixed wing ASW and fighter aircraft ... In general, the tactical advantages to be gained resulted from the helicopter's speed, versatility, and immunity from submarine counter-attack. In particular, the advantages lay in the capabilities of the aircraft as an extension of the ship's sonar system and the ship's weapon capability. Both were significant additions to the capability of the surface ship, giving it the initiative of attack and exposing the submarine to the threat of detection, classification, and attack without prior warning and at ranges exceeding the sonar and weapons systems of the ship.[77]

So, while it was true that Canada did not have much room aboard its carrier for helicopters, it was fully realized that there was a far more important advantage to adding the aircraft – an expansion of the eyes and ears of the Canadian ships.

Clearly, there was very judicious analysis being undertaken regarding the wedding of helicopter and warship in the Canadian navy during the late 1950s. The Naval Staff truly aspired to ascertain both present and potential effectiveness of helicopters in ASW and to configure a policy for their use in the RCN. Canada was, in fact, at the forefront of conducting trials with a helicopter equipped with both sonar and weapons.[78] More trials aboard HMCS *Ottawa* reaffirmed that a haul-down and securing system was needed, but they also confirmed that a hangar was essential for maintenance and shelter. An RCAF helicopter was given to the RCN on the understanding that it would be returned undamaged; however, it was sent back with such a severe case of saltwater corrosion that it needed a major overhaul and created tension between the services.[79] Notwithstanding this minor setback, the final report concerning the five weeks of exercises aboard the *Ottawa* was very positive and asserted that "it is entirely feasible to operate an HSS helicopter from a ST. LAURENT Class Escort in the North Atlantic for a worthwhile percentage of the time." It was recommended, therefore, that operational platforms be installed on all *St. Laurent/Restigouche* class escort vessels immediately.[80]

While the Project Definition Phase continued within the Naval Staff, the Policy and Projects Coordinating Committee (PPCC), and the Naval Board concerning the helicopter to be used, the helicopter-carrying escort concept had been formally accepted by 1959. In January, the Naval Board agreed that the fifth and final sixth *Repeat Restigouche* class destroyers that remained to be built would be more capable and would be fitted with a Variable Depth Sonar (VDS), a flight deck and hangar to protect a helicopter, an SQS 503 hull-mounted sonar to work with it, and guided missiles.[81] These ships would later be renamed the *Annapolis* class destroyers and were built specifically to carry a helicopter.

VDS was a new type of sonar created to deal with the fact that thermal stratification of water will bend a sound beam and allow a submarine to escape detection. The point where the temperature changed from the surface temperature curve to the deep-water temperature curve became known during the Second World War and was described as the "layer." VDS was designed in Canada to get around this problem.[82] The sonar head could be put on a cable and lowered and towed by the ship at varying depths until it was under the layer and able to detect concealed submarines. Although it was a relatively small equipment project, it was a source of particular pride for Canadian ASW and a testament to its focus on that area. It was so efficient that even the British chose to abort their own design attempts and procured the Canadian version in 1960.[83] By the end of the decade, VDS became vital to the RCN's Anti-Submarine Warfare (ASW) capability.[84]

Part of the impetus for these developments was that the Soviet navy was making major technological advances in submarine warfare in the late 1950s. The USN was working to counter this threat, and an American submarine called the *Nautilus* had actually travelled under the polar ice cap in 1957. Its sister ship, the United States Ship (USS) *Seawolf,* had actually remained submerged for sixty days in the Atlantic and covered 15,700 miles.[85] These developments were significant as they made submarine operations in Canada's northern waters viable. Furthermore, SOSUS had failed to provide a reliable means of detecting submarines for Canada. After a contact had been made off Shelburne, Nova Scotia, it took RCN ships five hours to reach the contact point, and even then they failed to find the submarine due to limited sonar ranges.[86] Surface vessels alone were never intended to counter SOSUS-generated targets.

George R. Pearkes had become the Minister of National Defence after the Conservatives won the election in 1957 and was anxious about Canadian ASW capabilities. He had previously stated that "modern submarines can come within a reasonable distance [of our shores] and from those submarines can be launched missiles or even aircraft."[87] To locate and track the submarines, he concluded, there should be the fullest possible use of reconnaissance planes and helicopters working with the fleet arm. There was even a Nuclear Submarine Survey Team approved by the Naval Board in 1958 to assess the feasibility of building ASW submarines in Canada.[88] Due to the excessive cost of the vessels, however, the focus remained on helicopters.

In September 1959, the Projects Committee was given a document titled "Staff Characteristics for an Escort Borne ASW Helicopter," and the ACNS informed the committee that emphasis had been placed on "simplicity of design with reliability of components, ease of maintenance, convenience of size, and reasonably low cost of providing an anti-submarine helicopter capable of being operated from platform fitted ships and aircraft carriers in all weather, by day and night."[89] Interestingly, the size of the aircraft was an issue. The problem was that, since the RCN was the only navy considering flying large helicopters from DEs, it had difficulty finding a capable aircraft of the right size. And since Canada did not plan on designing its own aircraft, the procurement process, especially the Project Definition Phase, took longer than anticipated. The RCN was essentially trying to create a platform before it was certain which weapon would be used within it. As the Canadian military had learned many times before, choices were limited when another nation was designing its weapons and equipment.[90]

Although a great deal of time had been spent on the Staff Requirements for an ASW helicopter, neither of the two frontrunners by that point, the Kaman *HU2K-1* and the Sikorsky *S-63,* had ever flown in a naval configuration, and neither met all the specifications of the RCN. The vice chiefs of staff recommended to the chiefs of staff later that September that they settle on the Kaman *HU2K-1* as it was closest to the requirement. The *HU2K-1* was designed to meet the SAR role in the USN and was smaller, faster, and lighter.[91] The chiefs of staff subsequently agreed, and the Deputy Minister of Defence negotiated with the DDP for the procurement of forty helicopters at a rate of six per year.[92] The RCN program, as outlined to the Deputy Minister in October 1959, scheduled the procurement for a period from 1961 to 1966.[93]

A procurement that was originally to take about five years – begun in 1954 and to be done by 1960 – was, by that point, a year or two behind. According to the ACNS, this situation seriously reduced the navy's long-term ASW capabilities. But notwithstanding that the project was beginning to fall behind schedule, there was still a real commitment to acquiring the new capability, and the reasons for its delay were valid; it was a new platform still in the trial phase, and nobody was certain which model would fit well into the new escort-helicopter ASW tandem. The entire concept was a significant gamble by the RCN, and it provided the necessary justification for the length of time spent on the Staff Characteristics. They were finalized in 1959. Overall, however, RCN officials originally expected to acquire the aircraft by 1961.[94]

As the Cold War heightened in the late 1950s, the potential of the Soviet nuclear submarine threatened to make existing naval fleets obsolete. Canada's allied obligations were firmly entrenched in ASW, and the Canadian navy continued to make its case for expanding detection and attack capabilities. The pressing concern was that, although the *St. Laurent* class vessels were entering service and plans for a new *Mackenzie* class were under way, the RCN still did not have enough ships for its international commitments. It was also concerned about obsolescence. But with a limited defence budget, the procurement of more advanced destroyers was simply not possible. The Naval Staff saw a solution in the naval helicopter and subsequently looked to it as a way to expand the ability of their ships then under construction. It became apparent after extensive analysis and experimentation that they could enable independent escorts and destroyers to carry a modern ASW helicopter rather than limiting them to the sole Canadian aircraft carrier. The previously amorphous role of the naval helicopter was taking shape in the RCN, and it became the global leader in the research and development of this new weapons platform.

By 1959, the helicopter-carrying escort concept had been formally accepted in Canada, and this acceptance allowed the procurement officials to better define their requirements. Actual acquisition remained elusive, however, as their vision surpassed the actual specifications of any existing helicopter; they could see the future, and they refused to recommend a purchase that would be outdated shortly thereafter. The key difficulty eventually encountered was that, since the RCN was the only navy considering flying helicopters from smaller ships, it had trouble finding a capable aircraft of the desired size. It was clear that Canadian industry could not produce the type of aircraft demanded, and the RCN could only wait for new designs from its allies, notably its most dominant source of helicopters, the United States.

Active Endeavour: Hoisting exercise aboard HMCS *Iroquois.* Petty Officer 2nd Class Brian Bishop is lowered down to the flight deck on 22 October 2006. Bishop, staff communications specialist, is a member of SNMG1 Staff.

Source: CF photo by M.Cpl. Charles Barber. *Courtesy of National Defence. Reproduced with the permission of the Minister of Public Works and Government Services, 2009*

3
The Procurement of the *Sea King*: Slow but Solid

> This new helicopter, when it becomes fully operational[,] will be employed at sea not only from the deck of Bonaventure, but by our Canadian destroyer escorts, which are being equipped with landing platforms ... I should like you now to take note of the fact that the concept of operating multi-purpose, large, sonar-equipped helicopters from destroyer escorts is a Canadian one, and I am confident it will prove to be an important contribution to the strength of our western defences.
>
> – Address by Rear Admiral Jeffry Brock on the Occasion of the Introduction of the *Sea King* Helicopter into the Royal Canadian Navy (RCN), 28 August 1963

The time of the Canadian naval fighter aircraft came to a close in the early 1960s, and the integration of the Anti-Submarine Warfare (ASW) helicopter into the RCN evolved into something truly unique to Canada.[1] On 9 June 1960, a submission was sent to the Treasury Board from the Naval Staff to demonstrate a planned program of improving the anti-submarine efficiency of the fleet by fitting more vessels with Variable Depth Sonar (VDS), long-range hull-mounted sonar, and, of course, helicopters. The modernization program sought approval for the conversion of Second World War-vintage *V* class Her Majesty's Canadian Ship (HMCS) *Sioux* and *C* class *Crusader* and the *St. Laurent* class vessels to include these new technologies. The Naval Board was hesitant to convert the latter class so soon after the ships had become operational. Any conversion would also reduce the number of operational ships committed to NATO.[2] But after the acquisition of nuclear submarine capabilities by the Soviet Union, the ship conversions were considered necessary to keep up the ASW capability of the RCN. And though the Treasury Board approved the modifications, they were still hesitant about selecting a helicopter for fear of obsolescence; they wanted to be sure they were getting the newest available model.[3]

On 27 January 1961, the Canadian Naval Staff presented an analysis paper to the Naval Board. The minutes show that the Kaman *HU2K-1* helicopter in particular was considered suitable for operations from the Destroyer Escorts (DEs) scheduled for conversion. The board referred to another model, the Sikorsky *S-61* (*HSS-2*), or what would later become known as the *Sea King,* only to state that it "had been ruled out as far as the RCN was concerned because there were still development difficulties; it was too heavy and large to meet the staff characteristics and had been estimated to cost 40 percent more than the *HU2K-1*. It was also a more complex machine to maintain."[4] The board was aware that the *HSS-2* had already been selected for carrier operations by the United States Navy (USN), and it was arguably the first helicopter designed expressly for naval applications. The aircraft was a five-blade main and tail rotor system that combined both hunter and killer capabilities: it could detect, identify, track, and destroy aggressor submarines. The *HU2K-1* was more of a multi-purpose naval utility model for Search and Rescue (SAR) and passenger transport. But since the RCN was considering using helicopters on smaller ships, the *HSS-2* was initially overlooked.

Since it was admitted that the Kaman *HU2K-1* did not meet all the Naval Staff Requirements, it was agreed that only twelve would be procured in the initial order and that no firm requirement would be set for the type that would follow the initial supply.[5] The *HO4S-3* would have to fulfill any interim role. A Treasury Board memorandum dated 23 June 1960 had revealed that

> the Board agreed in principle with the overall concept of the programme but felt that the procurement of helicopters at this time may be premature. It was understood that the Kaman helicopter might require modifications to adapt it for use on the RCN role. For this reason and in view of the current rapid changes in the state of the art, the Board considered that it would be preferable to defer this portion of the programme in the hope that an aircraft would be available in the near future for which there would be less risk of obsolescence than for something available at this time ... It was suggested that these helicopters [*HO4S*] fill any time gap which might arise between conversion of the first ST. LAURENT vessels and receipt of new production aircraft and would give the Crown some flexibility in timing the major move into the helicopter field. This is desirable to avoid having a further major purchase come up too soon on the grounds that rapid developments had made these Kaman helicopters obsolete.[6]

Although this was a sound judgment by the Treasury Board, military technology is always evolving, and a government, in conjunction with its military, must decide to enter the field at some point; the procurement of a new

ASW helicopter to correspond to Canada's extensive testing of helicopter-carrying escorts had been discussed since 1954 and was supposed to have been accomplished already. Without these helicopters, the newly converted ships would be impotent. This situation was exacerbated by the fact that SOSUS was still unable to reliably detect Soviet submarines. In November 1961, Vice Chief of the Naval Staff Jeffry Brock informed the navy's senior officers that the RCN had no reliable ability to stop missile-firing submarines.[7] Brock had already submitted a report that summer on a replacement fleet for the RCN, and it was debated by the Naval Board between September and December. The Brock Report called for a replacement and modernization program that included six *Barbel* class conventional submarines, replenishment vessels to increase operational endurance, eight General Purpose Frigates (GPFs) to replace the Second World War *Tribal* class destroyers, and the construction of "heliporters" to carry helicopters in the ASW role. That the Naval Board endorsed the report and approved its submission to the Minister of Defence and Chiefs of Staff Committee demonstrated that they all supported the immediate introduction of helicopters into the fleet and that ASW would continue to be the focus of the RCN. Unlike most other NATO navies, the RCN would not restructure around carrier task groups or strike forces.[8]

In a letter from the Deputy Minister of National Defence to the Assistant Secretary of the Treasury Board, the former asserted that

> I am satisfied that the point in time has now been reached to choose the ASW helicopter and that the Navy has conducted extensive and thorough helicopter evaluations to meet the operational requirement. An alternative to selecting an existing helicopter would have been to launch a design competition to industry for an ASW model designed to meet the Navy's escort based concept: the risks attached to this approach would be out of proportion to the advantages which might be obtained.[9]

Canadian industry was in no position to create a helicopter and have it manufactured domestically to meet the desired schedule for delivery.

As a result of the Deputy Minister's letter, the Treasury Board amended the 1960-61 procurement program of the RCN on 5 December 1960 and approved procurement of the twelve *HU2K-1*s at an estimated cost of $14.5 million. Operational trials would determine whether the *HU2K-1* was suitable for an RCN need of forty ASW helicopters, delivered at a rate in step with ship conversions. Although this approval seemed to conclude the matter for the RCN, new developments stalled the proceedings once again. It was discovered that "subsequent preliminary contract negotiations between the Department of Defence Production (DDP) and the Kaman Aircraft

Corporation had disclosed that the procurement cost had increased to $23 million or an additional cost of almost $2 million for each aircraft."[10] Kaman's decision to raise the price was not disclosed to Canada before this, and the DDP wrote to the Deputy Minister in frustration, describing how the figures "bore no resemblance to those which had been proposed."[11] The Kaman *HU2K-1* had been picked originally because of its "cheapness."[12]

The new figure, in addition to the fact that the Kaman model required modifications to embody an ASW capability, caused a major reversal of opinion among the Naval Board. There had actually been some dissension over the purchase from the beginning. As early as December 1959, the Assistant Chief of the Naval Staff (ACNS) had stated that, though he was behind the procurement of the Kaman model in the interest of time, that model had "no growth potential for the future." He continued by saying that "it is annoying that for the expenditure of $16 million we shall still not have *everything* we want in a helicopter."[13] Indeed, the Kaman had limited endurance and could carry only a small payload relative to other models. It was, after all, designed as a utility helicopter. With a high degree of prescience, Commodore Brock concluded that, "until our final decision is made and an order placed, there are bound to be new and interesting proposals made at frequent intervals ... It is certain that a better machine than Kaman could be produced if we are willing to wait for it. It is probable that a better machine will be produced in the not too distant future." Likely, he was even less interested in the Kaman once he discovered that the price had risen by 30 percent.

After the price increase of the *HU2K-1,* the Naval Board looked at its other options once again. In January 1961, they concluded that,

> In view of the large price increase of the HU2K, it was considered that the USN choice, the [Sikorsky] HSS2, would provide better value for the money if the larger helicopter could be suitably accommodated in RCN ASW escorts. NCC [Naval Constructor-in-Chief] had re-examined the possibility of accommodating the HSS2 and considered that this could be done with certain modifications to the ships. The aircraft could now be obtained as a true "off-the-shelf" item and, with a recent price decrease, would be within the general cost parameters of the HU2K. Though previously dismissed because of cost, size, and complexity, it now appeared that the only major disadvantage was size.[14]

Other models were subsequently analyzed to be procured in lieu of the *HU2K,* such as the Vertol *107,* the Westland *Wessex (S-58),* and the Sikorsky *S-65.* But it was the Sikorsky *HSS-2* that became the new favourite for the replacement program.[15] As a result, it became necessary to determine what the modifications would be to RCN vessels in order to accommodate the larger aircraft. The Assistant Naval Constructor-in-Chief (ANCC) reviewed the alterations that would have to be undertaken for the DEs and concluded that

> For the 265 class this involved splitting the existing funnel uptake in two; moving the existing hangar as far forward as possible; extending the after end of the flight deck aft and general strengthening to take the increasing landing blow. An additional 30 tons of extra structure would be required which would make stability in the light condition marginal and the ship might have to employ water ballasting. The arrangements would be approximately the same for the 205 class modernized.[16]

The modification costs for each of the *265* class were calculated at $85,000 to $100,000 and for each of the *205* class between $75,000 and $90,000.

Although members of the Naval Board were hesitant regarding the extra costs for ship modifications that the *HSS-2* required, it was eventually decided to stick with it as new information became available. In August, it had been discovered that trials with the *HU2K* by the USN had revealed serious problems with the aircraft design.[17] At the same time, however, the Naval Board revealed that *HSS-2* production was ahead of schedule, with the first eight aircraft having already been delivered to the USN. In addition, because of quantity production, the unit price had also been reduced.[18] By that time, the helicopter was officially called the *Sea King*. It was a more suitable aircraft for HS50 and for carrier operations; it could be considered for future replacement of the CS2F *Tracker;* it had better growth potential; it could carry more troops than the *HU2K;* it could be flown on one engine, if necessary; it could carry both weapons and sonar; and, something that was vital to the RCN's smaller ships, it had power folding rotor blades. The *Sea King* was also used by the USN for ASW, and this meant a certain degree of standardization of equipment that was so popular at the time. It had entered service in the USN in June 1961, and since it could be purchased by Canada "off the shelf" this would result in "common user benefits," such as accessibility of spare parts.[19]

The ACNS submitted to the Naval Policy Co-Ordinating Committee (NPCC) that the enlargement of the helicopter facilities in the *St. Laurent* class be "proceeded with immediately to enable an HSS2 size helicopter to be embarked and operated."[20] He also desired that authority be granted for an RCN team to visit the Bureau of Weapons and the Sikorsky Aircraft Corporation to examine and report within thirty days on the operational, technical, and financial considerations involved in selecting the *HSS-2* as an alternative choice to the *HU2K*. Investigation had also shown that existing water ballast tanks could provide adequate compensation for the additional weight of the *HSS-2*. The *HSS-2* consumed more fuel than the *HU2K-1,* but it was thought that increased stowage could be found. An increase in maintenance personnel to work on the more complicated aircraft was also accepted as part of the plan. As the Naval Staff within the NPCC concluded, "the Sikorsky HSS-2 shows great promise of being the most effective vehicle of its kind."[21]

The decision on the procurement of the *HSS-2* was to be made once it was finally determined whether the necessary facilities could be provided aboard the RCN DEs. In fact, a submission to the Treasury Board had already been prepared for the procurement of the *HSS-2*. It was noted at the meeting that ship stabilization was not essential to helicopter operation from escorts; however, stabilization would decrease significantly the amount of flying time that would be lost in poor weather conditions. The cost of providing stabilization, it was believed, could be absorbed in the cost of the conversion of the ships. Plans were already being prepared for suitable hangars and flight facilities, and the Director General Aircraft stated that modification of the tail pylon folding arrangements, which existed to create more room on the flight deck, from manual to hydraulic would present no serious problem. Significantly, the Naval Board did not consider sea trials necessary as they deemed that the trials aboard HMCS *Ottawa* had already proven the concept of operating an ASW helicopter from an escort vessel. Furthermore, the *HSS-2* was judged to be "superior in all respects" to the first *HSS* model (*H-34/S-58*), with which the *Ottawa* trials had been conducted. The schedule, therefore, called for commencing ship conversions to accommodate the *HSS-2* in mid-1962 and to complete them approximately a year later.

Part of the ship conversion necessary for the RCN was the production, testing, and installation of the Canadian-designed *Beartrap* Hauldown and Rapid Securing Device. The possibility of procuring the *HSS-2* helicopter aboard its frigate/destroyer-size warships added impetus to the successful creation of a type of cable system for bringing a helicopter down securely onto the ship. Previous trials aboard *Buckingham* and *Ottawa* had demonstrated that this was likely the best option for landing and securing the helicopter in the hangar. As already noted, VX 10 also had some success pulling a helicopter down by a tensioned wire. Lieutenant Commander Frank Willis from the Staff of the Director of Canadian Maritime Forces Operational Requirements Air explained in 1962 how the problem of wedding the helicopter to the ship was "tremendous." The two most important conclusions reached at the time were these:

> The platform must be located as high above the water line as possible to reduce the effects of salt spray on the windscreen and on the fuselage ... The other requirement is for a method of bringing the helicopter down to the deck as rapidly as possible and holding it secured to the platform firmly at all times when it is on the deck. The hauldown, beartrap and traverse system were subsequently designed by our Naval design staff.[22]

Although the new maritime helicopter was clearly going to be purchased from the United States, some members of the Treasury Board wished to explore Canadian assembly of the aircraft to include "Canadian content." They were fully aware that it would be too costly and time consuming to hold a helicopter design competition for strictly Canadian needs; however, if Canada were involved in the assembly, then at least it could stimulate some aspect of Canadian industry. And there would be a possibility of modifying certain aspects of the helicopter to fit the Canadian navy's specific requirements; the RCN was, after all, creating a unique weapons platform. But the DDP had advised the Treasury Board that there would be substantial cost increases involved in production in Canada. After studying the matter, the DDP sent a letter to the board outlining its conclusions:

> Because of the small number of the immediate requirement (ten) and the long lead time on the additional potential requirement for 26 machines, it would not be economically sound to set up airframe production in Canada. The premium for total production which would necessitate a complete set of tooling, test equipment, license agreement, excess early production costs and a re-arrangement of the production facilities at Canadian Pratt and Whitney Aircraft Company Limited, Longueil, Quebec, and associated industries amount to over $7 million.[23]

The "excess early production costs" alone were estimated to amount to $120,000 per aircraft for approximately 20 percent Canadian content in the final model. The letter continued: "It is not recommended, therefore, that any degree of production of the airframe in Canada be undertaken." There was also discussion about building the engine, the General Electric T-58, domestically. The DDP was equally adamant that not enough engines were needed and that the tooling and licensing costs would be too high. As always with the purchase of another nation's design, if Canada wanted to help in the production of the final model, it would have to pay and then likely wait for permission from the company. Those at the DDP were not interested in either of these entanglements. They were not alone. An interim report on the *Royal Commission on Government Organization* in 1962 headed by Air Vice Marshal F.S. McGill, commonly called the McGill Report, stated that "the residual activity of ensuring contracts are given to Canadian industry has imposed long delays and additional costs over which DND [Department of National Defence] has no control ... To have to effect design changes and modifications through an intermediary is both time-consuming and costly."[24] Furthermore, the report claimed that the purchasing aspect of the DDP was already leading to "duplication and overlapping with competent sections in DND where the necessary screening, reviewing, and consolidation of requirements [are] done as part of the process of preparing Contract Demands." The report recommended the creation of a Deputy Minister (Procurement) within the DND to replace the DDP role of buffer between the military and the contractor.

In the response to the DDP's letter regarding the flaws in helicopter production in Canada, Assistant Secretary of the Treasury Board J. MacNeill wrote that

> The Board noted your conclusion that, because of the small number of the immediate requirement and the long leadtime on the additional potential requirement, it would not be economically sound to set up airframe production in Canada. The Ministers were not convinced, however, that another alternative which you do not recommend does not hold reasonable possibilities at an acceptable premium; that is, the alternative of partial production by means of assembly in Canada of components and equipments imported from Sikorsky, with a 20 percent Canadian content. While the premium had been estimated at $120,000 [for] each of the first 10 helicopters ... the Board is extremely interested in having this premium recalculated on the basis of the full RCN requirement.[25]

Although MacNeill ended his communication by asking that Golden reconsider the possibilities for airframe assembly of the *HSS-2* in Canada, he did write that the Treasury Board was satisfied with the recommendation to purchase engines from the United States "in view of experience to date."

Whether or not there would be any Canadian content aboard the aircraft, there was still no platform big enough in the RCN to operate the *HSS-2* from the back of the ship. After all, the aircraft had not been designed to operate on anything smaller than a carrier. A memo to the Deputy Minister of National Defence, however, revealed that the first permanent platform, with a 60 percent increase in size, would become available in late 1963.[26] Once the decision to procure the *HSS-2* was near finalization, an extensive study was done on landing the helicopter aboard its new shipborne base of operations. Three classes of vessel were listed to carry the helicopter: the *St. Laurent,* the *Restigouche,* and the new *Mackenzie/Annapolis (Repeat Restigouche).* In each class, the landing platform was to be situated in the after end of the ship. Although the Naval Staff were confident that the *HSS-2* would function well with the escorts, they eventually decided that it would be best to borrow one of the American models to test the stability of the aircraft in a haul-down demonstration. Although the first ten helicopters procured were to go to the carrier-based squadron, HS50, this was to be the final test to prove the viability of operating the *HSS-2* on DEs before the government committed to Sikorsky. The United States agreed to loan to Canada a USN helicopter subject to a guarantee of indemnity and a payment based on hours flown.[27] There was also the option of buying one prototype and using it for the test. Vice Admiral Rayner rejected this option, however, as he did not even think that this final test was necessary. Experienced RCN pilots also did not believe this to be "an economical course of action."[28] Rayner also stressed that action had to be taken quickly as "the helicopter replacement program is already well behind schedule owing to Kaman's failure." The Minister of National Defence concurred with this conclusion. Notwithstanding protest, the trial was necessary to conclusively prove to the Treasury Board the viability of the system, and it was carried out at the Sikorsky plant in Stratford, Connecticut, on 16 April 1962. It was a complete success.[29]

On 20 November 1962, Minister of National Defence Douglas Harkness formally announced approval to procure the Sikorsky *HSS-2* for the RCN.[30] The stated requirement was for an immediate procurement of eight aircraft, with an ultimate purchase of forty-four.[31] Harkness stated that negotiations to acquire the eight aircraft for the RCN in 1963-64 were under way but explained that, "due to the long production time required for some of their weapons systems, the first three aircraft will not be fully equipped until 1964."[32] Until then, the helicopters that were received were to be used in a training role. In fact, Sikorsky had already offered to train Canadian maintenance crews in the United States before the first three aircraft were delivered since training aids would not be available in Canada in time for training personnel for the first three aircraft and were "unlikely to be available for training personnel for the further five aircraft."[33] Sikorsky offered to help train pilots as well, but there would be a cost involved, and the number would be limited to as many as the company thought it could take on in addition to training its own pilots.

Once it was clear that there would be approximately forty-four helicopters procured, thoughts returned to the possibility of assembly in Canada. It was recommended to the Deputy Minister of National Defence by the Director of the Aircraft Branch in the DDP that the Canadian government buy the first three helicopters directly from Sikorsky but pay for the necessary tooling and test equipment to allow the remainder of the machines to be assembled in Canada. An estimate of the additional cost to do so was included in a letter to the Deputy Minister:

Tooling	$400,000.00
Test and Support Equipment (for production only)	$300,000.00
Contractor Plant re-arrangement	$150,000.00
	$850,000.00
Estimated excess early production cost	$250,000.00
Estimated Total	$1,100,000.00

This total amortized over the eight aircraft amounted to approximately $137,000 per aircraft. It was pointed out to the Deputy Minister that the "obvious advantage in the larger program of forty-four machines is the lower price per unit resulting from a broader amortization base and economies in the production of larger numbers ... It is our [DDP] objective to obtain the greatest amount of Canadian content possible."[34] The Minister of National Defence agreed: "A programme of this size offers, at some premium in cost, a fully practicable opportunity for Canadian production and one that could be readily and economically extended to provide such greater numbers as may be required eventually."[35] The Canadian government subsequently decided to pay more for each model in order to use military procurement to stimulate domestic industry. The first four helicopters were subsequently manufactured and delivered by Sikorsky Aircraft, and the remaining aircraft were to be assembled at the United Aircraft of Canada plant in Montreal.[36] This arrangement went against the original recommendations of the DDP and many senior military officials, specifically Air Vice Marshal McGill, as outlined in his *Royal Commission on Government Organization.*

After the decision was made to procure the *Sea King,* the RCN made a request through the Canadian Embassy in Washington to have a USN model flown to Shearwater for demonstrations. Glenn Cook was asked by his Commanding Officer on the United States Ship (USS) *Intrepid* aircraft carrier if he would make a flight to Shearwater to complete the request. He had been serving in the United States since spring 1962 and had gained some experience in the USN with its *Sea Kings*. That March Cook arrived in Shearwater with the *Sea King,* and Canadian helicopter pilots were given a glimpse of the aircraft that would soon form the basis for ASW in the air.[37]

Although the *Sea King* project was near first delivery, the GPF procurement proposed by the Brock Report in 1961 was not so fortunate. Cabinet had agreed to the construction in Canadian shipyards on 19 March 1962. Michael Hennessy has written that "after nearly four years of consideration the ship replacement programme was signed and sealed. Delivery proved another matter."[38] One of the problems was that the industrial design capacity created for the *St. Laurent* and *Restigouche* programs during the early 1950s was mostly non-existent by the time the GPF program was under way; the skilled engineers and draftsmen had moved on. The DDP was also deficient in technical staff or experience necessary for risk assessment and cost control. As a result, there were recurring problems with the design, and costs quickly escalated beyond what was projected by the department. Moreover, the RCN had chosen the USN's *Tartar* surface-to-air missile system when it was still under development, and it was behind schedule. Ship construction was to have begun before the end of 1963, but it was set back to July 1964. The cost of the vessel had also risen from $33 million per ship to $36.2 million. This was partly because the Naval Technical Services had relied simply on the manufacturer's approximate figures that were completely insufficient for completing the designs. After Lester Pearson and the Liberals were elected in 1963, they quickly balked at the runaway project based on concerns by the Treasury Board and the Department of Finance. After a government project review, the new Minister of National Defence, Paul Hellyer, rose in the Canadian House of Commons on 25 October 1963 to announce the cancellation of the GPF. Hennessy has concluded that the cancellation of the program was more about the Liberal plan to cut the overall defence budget and generally reform defence policy. Although this conclusion may be partly correct, the extensive scheduling delays, lack of design finalization, and massive cost overruns made the GPF an obvious and easy target for a new government. Hennessy has also noted that the cancellation meant that twenty-three surface vessels built during the Second World War remained in service for years past their effective lives. They all lacked modern ASW equipment.

On 28 August 1963, Rear Admiral Jeffry Brock spoke on the official occasion of the introduction of the HSS-2 *Sea King* helicopter into the RCN. He confidently stated that

> The introduction of the Sea King helicopter into the Royal Canadian Navy represents an advance of some considerable consequence in the field of anti-submarine operations. This new helicopter, when it becomes fully operational, will be employed at sea not only from the deck of Bonaventure but by our Canadian destroyer escorts, which are being equipped with landing platforms. The Sea King will extend considerably both the detection and destruction capabilities of our anti-submarine ships. Most important, it will help to eliminate the speed advantage hitherto enjoyed by the nuclear-powered submarine.[39]

Brock stressed that it was more capable and versatile than any previous helicopter used by the RCN not only for ASW but for search and rescue as well. Canada was a recognized expert in the field of ASW, he said, and pointed with pride to the Canadian development of VDS that was being adopted by other countries and to the fact that flying ASW helicopters aboard DEs was uniquely Canadian. Brigadier General Colin Curleigh, one of the pilots for the delivery of the initial Canadian *Sea Kings* from the Sikorsky plant and later Commander of Maritime Air Group from 1986 to 1989, explained how original the idea was at the time:

> The RN [Royal Navy] initially operated the diminutive *Wasp* from some of its frigates, while the USN took the unmanned route with its ill-fated Drone Anti-Submarine Helicopter ... The RCN embarked on a much more ambitious

> approach ... The momentous decision to investigate using the large *Sea King,* which was already being flown off the Bonaventure, quickly followed. Other navies thought we were crazy, and there were moments when we thought they could be right.[40]

And so began *Sea King* operations in Canada.

In 1963, Vice Admiral H.S. Rayner, Chief of the Naval Staff, spoke to a Special Senate Committee on Defence to discuss the purpose of the navy, its strength, and naval expenditures in recent years. The first thing he spoke of was the need to defend against the Soviet maritime threat. He explained how the Soviet fleet included over 400 submarines, of which over twenty were nuclear powered, and that they were "reaching out over the oceans."[41] This assessment may seem melodramatic in the post-Cold War era, but it was a real concern at the time and certainly justified expenditure on ASW warfare in Canada, with its vast coastal approaches and high level of overseas trade. Rayner also spoke of international commitments of collective security within NATO and how Canada had been given an integral role in ASW and how it was important to watch over Canadian sovereignty in the Arctic. As Rayner discussed the capabilities of nuclear submarines, such as their speed and potential for being submerged "for months on end," he came to the rationalization of ASW helicopters. He explained that they could effectively track submarines and that their speed could overcome even the most modern version. When they were used as an integrated system with a host ship's sonar, helicopters were deemed essential for relocation, classification, and attack on submarines. He concluded that "the challenge for us is to ensure that our country ... will have a strong navy in the years to come. We will do our utmost to meet this challenge but we need the blessing and firm support from the people of Canada." Clearly, Rayner understood the political nature of equipment procurement in Canada and was determined to convince his political masters of the necessity of continued investment in ASW capabilities. Even though three of the *Sea King* helicopters had been delivered by the time of this speech, Rayner knew that it was important to keep pressing the point home. At the Senior Naval Officers Conference the following year, the Minister of National Defence revealed a similar awareness: "The RCN must not only continue to improve its anti-submarine capability, but impress the public with the Navy's ability in this field to obtain greater public support."[42]

Although political considerations dominate weapons procurement in Canada, there have been occasions when the government has taken the advice of its military advisers and acquired a piece of equipment based primarily on the capabilities that it would deliver to the Canadian Forces. The procurement of the Sikorsky *Sea King* was based on compatibility, capability, and cost and was carried out because it was the best platform for what the military was trying to undertake; political considerations did not dominate the procurement. Although the need for ASW capability stemmed from Canada's political commitment to its allies, the acquisition was based on finding the best helicopter to carry out this mandate. The Naval Staff had envisioned the introduction of a modern naval helicopter to fly from its DEs throughout the 1950s, and the *Sea King* was the only model that could fulfill their discerning and demanding requirements. Brigadier General Curleigh later recalled that "as we proudly delivered our shiny new helicopters we were hailed as part of the first step in initiating a new era in maritime aviation and naval capability."[43] The purchase of the most modern equipment for the navy was not an easy task, however, as the desire to procure new equipment for the military in the early 1960s was broadly undermined by the weak fiscal position of the government. This pecuniary position was partly overcome by the success of the government and its military advisers through education of the citizenry on the naval threat to North America and Canada's commitment to NATO to meet it. The pervasive fear of Soviet expansionism and technological capability by the 1960s made the purchase of a modern ASW helicopter a convincing argument to Canadians and, therefore, to their political leaders.

Although it is obvious, in retrospect, that the *Sea King* was the best option for the RCN, the decision to purchase it came only after exhaustive analysis, introspection, and re-evaluation. It also had to await many developments in the United States before it could be carried through. Indeed, two other helicopters had been chosen in succession to fulfill the naval helicopter role ahead of the *Sea King*. As the procurement process was carried out, however, the *Sea King* slowly revealed itself to be vastly superior to the other models in contention. It had initially been overlooked as a possibility mostly because of its massive size. But once it was determined that the necessary technology could be designed and produced in Canada to allow it to land and be secured on the ships, the reversal followed quickly. The unique operational role that helicopters would play in Canada demanded modifications to the American design. The Canadian government also insisted that most of them be assembled in Canada. As always, the reality of procuring a weapon platform designed for the needs of another nation caused major delays in its introduction. Nonetheless, by July 1963, the first four helicopters had been accepted by the RCN.

HMCS *St John's:* Cpl. Dan Fanning, an Aviation Technician (AVN), cleans and checks pylon on a CH-124 *Sea King* helicopter before take-off for the first mission of the day on 15 November 2007.

Source: CF photo by M.Cpl. Eduardo Mora Pineda. *Courtesy of National Defence. Reproduced with the permission of the Minister of Public Works and Government Services, 2009*

4
The *Sea King* in Canada: Time Is the Enemy of Us All

> Aircraft production will be scheduled as nearly as possible to meet the operational, training, attrition, and pipeline requirements ... This production of a total of 43 helicopters is calculated to support a programme until 1975, at which time the useful life of the Sea King is expected to expire.
>
> – Naval Secret Staff (NSS), 20 May 1964

Once the initial three *Sea King* helicopters were received in Canada in the summer of 1963, there remained the problem of transforming them into fully operational Anti-Submarine Warfare (ASW) helicopters and completing flight and maintenance training of Royal Canadian Navy (RCN) personnel. The aircraft were designated as CHSS-2 *Sea Kings*. It was originally believed by the Naval Policy Co-Ordinating Committee (NPCC) that there would not be sufficient pilots or training until the spring of 1966.[1] There was discussion on intensifying the use of United States Navy (USN) training facilities at Key West, but doing so was exceedingly expensive, and there were limitations on the numbers that the USN could accept. But since there were no training aids, weapons systems trainers, or flight simulators in the RCN, there was little choice. The cost to train one pilot in the United States was $96,000, and funds were not available in the 1963-64 budget. One solution was to acquire a US version of the *Sea King* and fit it with RCN equipment for operational training, but it was also determined to be decidedly expensive. Although "type-training" could be done on an American model, all operational training logically had to be done in the Canadian version. The Canadian *Sea King*'s subsurface acoustic detection equipment differed from the US model, and pilots had to learn how to land on a Destroyer Escort (DE) using the *Beartrap*. The Canadian version also included automatic tail pylon folding, a fuel-dumping system, and a strengthened undercarriage.[2] There was also the Canadian addition of the distinctive foreign object damage deflector (FOD MOD) in front of the engine intakes designed to prevent the ingestion of ice.[3] Even if the Canadian *Sea King* was used in a US training facility, there was the problem of how to support it while on a foreign base. It was clear that "US maintenance facilities might not be available and some equipment in the Canadian aircraft would be peculiar to the RCN." By 1963, many American *Sea Kings* had already been modernized or altered from the prototype on which the *CHSS-2* was based.[4]

Another problem in introducing the *Sea King* into the RCN was the wedding of the aircraft to the navy's new platforms. As the helicopters were approaching readiness, so too were the *St. Laurent* DEs from which they would fly. A helicopter-carrying DE was expected to be able to detect, identify, and destroy all types of submarines. In addition, it needed to possess sufficient growth potential to enable at least the next generation of rotary wing aircraft to operate from its deck. Although the new ships were to become operational at approximately the same time as the helicopters, it was believed that "the useful life of a ship may be considered to be about twice that of an aircraft."[5] As early as 1963, those involved in Canadian naval aviation were cognizant that it would not be long before newer and more advanced models appeared. This was only the beginning of ASW helicopters, and the RCN wanted to be ready to integrate them when it was time to replace the *CHSS-2*.

DE conversions on the *St. Laurent* class were mostly complete, but trials on the haul-down gear were not expected to begin until October 1963, and they were to take approximately six months. This entailed the most crucial and, indeed, the most exciting portion of any Canadian content to be included in the final model. The Director of Naval Aircraft Requirements (DNAR) explained that "allowing for the normal lead times for the production of this equipment, the haul-down gear would not be available in sufficient numbers for all the destroyer platforms until the Spring of 1966. Steps could be taken to advance these equipments [sic] availability but this would mean taking short-cuts which might prove unsatisfactory later and produce additional costs in the programme."[6] In the ensuing discussion, it was noted that the haul-down gear was necessary for the operation of helicopters from destroyers under all operating conditions. The landing of the aircraft was not the deciding factor, it was revealed; rather, it was the handling of the aircraft on deck. It needed to be secure *after* landing, and it was simply too heavy to move with manpower. It was further explained that, although the acquisition of the equipment could be accelerated to carry out the necessary sea trials, it would mean fitting it without proper testing at sea. The date of commencement for sustained operations of the *Sea King* was, therefore, in doubt in early 1963 due to a lack of overall program coordination concerning the actual aircraft acquisition, ship conversions, and training. This situation was exacerbated by the need to consult and pay the United States regarding these matters.

The sea trials of the *Sea King* actually began in October 1963 on board the *St. Laurent* class *Assiniboine,* and by February they were progressing on schedule, with upward of 120 landings and take-offs being conducted. The *Assiniboine* had been recommissioned in June following its conversion to a helicopter-carrying destroyer (DDH) with the removal of one of the Limbo ASW mortars and the aft gun mounting to make room for the aircraft hangar and landing deck as well as the twinning of the single boiler funnel. Also added to the ship were a Variable Depth Sonar (VDS), the *Beartrap,* and the ship stabilizers to compensate for the increased weight of the *Sea King.*[7] At a senior officers' conference at Naval Headquarters in Ottawa that February, it was stated that, though there were many development problems encountered in proving the system, the concept "devised by the RCN ... appear[ed] to be the correct one."[8]

The *Beartrap* haul-down system was being manufactured by Fairey Aviation Limited in Dartmouth, Nova Scotia, after the company was awarded the contract for a prototype in September 1962. The design department at Fairey Aviation had been responsible for converting the RCN's fixed-wing *Avenger*s into an ASW configuration. Dowty of Canada was responsible for designing and producing the winching segment of the system. The first version of the *Beartrap* was placed aboard *Assiniboine* in October 1963 in Esquimalt, and handling trials were carried out using a "dummy helicopter" built out of steel and concrete to roughly the same proportions as the *Sea King*. This was done to ensure that an aircraft of that size could be handled in the air and then on the deck without the undercarriage being torn off. Peter Charlton of Experimentation Squadron Ten (VX 10) recalled the event:

> As might have been expected, nothing would work. Mechanically, everything seemed fine. But nothing ran. Investigation revealed that the two 140 core colour coded electrical cables that ran from either control position ... ended by being connected to terminals in a junction box in what could only be described as a random manner. Enquiries within the Dockyard led to the fact that the job had been done by a civilian electrician of considerable seniority who was well known to be colour blind! Once we got this sorted out, we were off to the races.[9]

But the *Beartrap* also failed its first trials due to slow operation of its braking system and an inability to lock satisfactorily around the probe built onto the helicopter. A document entitled "Landing the CHSS-2 on the DDE" explained the landing process in its entirety:

> For landing, the helicopter hovers approximately eighty feet above the flight deck and over the reference landing position. A light cable is then lowered to the deck where it is attached to the haul-down cable. The light cable is then reeled into the helicopter bringing with it the haul-down cable, the latter unreeling through the trap in the flight deck. At the helicopter the haul-down cable attaches automatically to a probe protruding from the keel. The flight deck controller then starts the haul-down winch which pulls the helicopter to the flight deck where the probe of the helicopter is directed into the trap by the haul-down cable. The flight deck controller then activates the rails which lock the helicopter into position on and to the flight deck.[10]

The cable itself also broke during landings when higher tension was applied, and the winch control system had to be revamped. The deck-locking design itself was subsequently improved and, when retested, performed satisfactorily.[11] The reaction time of the winch control system also showed that a shock absorber was necessary to ease the "snatch" when the cable was first connected to the helicopter and tension was applied. Perhaps most importantly, a way had to be discovered to allow the aircraft to release itself from the *Beartrap* at any time. The original concept was to release the entire probe from the bottom of the helicopter; this idea had to be re-evaluated after the first prototype was built and engineers realized that it was not wise to release a 20-pound projectile under 4,000 pounds of cabled tension. A new housing for the probe was subsequently designed by Charlton as head of the technical office in VX 10; means were then provided to release the cable without the heavy probe.[12] The subsequent prototypes were then delivered to *Assiniboine* before it set sail from Esquimalt to Halifax, where more handling trials would be conducted and modifications made. On 23 November 1963, the first *Sea King* in Canada was hoisted aboard the *Assiniboine* with the help of a crane for those on board to assess whether it would actually fit in the hangar; although it was a very close fit, the placement was a success, and the ship set out to conduct its sea trials.[13]

The first landing of a *Sea King* aboard a Canadian DDH occurred on 27 November 1963. The first actual use of the *Beartrap* was made on 3 December by Joe Sosnkowski and Larry Zbitnew. A set of "grounding tongs" was used at the time to discharge the aircraft's static electricity, which was channelled through the messenger cable that descended from the helicopter and was used to connect to the haul-down cable itself. This lesson had just been learned as the first attempts to do so ashore resulted in the Leading Seaman being knocked right off his feet by the charge after he grabbed the cable. Although this could have led to tragedy had they been at sea, the sailor was unharmed, and the grounding wire on the tongs prevented any further incident.[14]

As the helicopter was directed into the four-foot-square trap, the locking rails would ensure that it was secure, and that it would be ready to be straightened and moved directly into the hangar after the rotor blades and tail pylon were folded. The whole process was under the direction of the Landing Safety Officer (LSO). The initial amount approved for the *Beartrap* project was $100,000. Once testing revealed weaknesses in the original design, modifications were necessary, and it was obvious that more money would be needed. It was quickly pointed out at the time, however, that "this development differs from many others in that an idea had to be conceived and made to work i.e. an invention was necessary, whereas the majority of development projects are, in a large measure, modifications, adaptations or new applications of existing ideas."[15] Indeed, the Canadian *Beartrap* was an invention that was to permanently alter ASW warfare. Faults in its early production and additional costs were, therefore, to be expected.

By February 1964, the problems with the original *Beartrap* design had been deemed solvable, and the project went forward as planned.[16] VX 10 continued trials.[17] The RCN was then in possession of its first four helicopters, and the first Canadian-built *Sea King* was accepted from United Aircraft in Montreal in May 1964.[18] The remainder of the forty aircraft planned for were to come at a rate of one per month commencing in July 1964. These aircraft were calculated to support a helicopter program until 1975, "at which time the useful life of the Sea King is expected to expire."[19] When discussing the future, it was determined that the period from 1973 to 1975 was the best time to begin procuring more helicopters, as new ships would be commissioned and ready to receive aircraft. The Naval Staff had already begun studies to determine possible characteristics of the vehicle that would be best to procure for the 1975-85 period, assuming that they would have the same number of helicopter-carrying ships.[20]

In the 1960s, Canada's Cold War policies were being reviewed. The high cost of weapons technology in the nuclear age had prevented timely equipment replacements and new purchases by the government. Canada was also in a recession. New procurements, however, were essential for the military to keep pace with modern weapons developments. Modern equipment acquisitions were also required to enable the Canadian government to honour commitments made within the North Atlantic Treaty Organization (NATO) and the North American Air Defence Command (NORAD), as the threat of long-range Soviet bombers and intercontinental ballistic missiles brought the potential for war to the continent of North America. This threat had been epitomized by the Cuban Missile Crisis of 1962. Although the threat was valid and clearly identifiable, investment in defence was under heavy scrutiny in the 1960s. By 1964, Paul Hellyer, Minister of National Defence, observed that "the government was very much aware of the overall financial position which faces it. A series of large deficits have increased the size of our national debt and the annual cost of servicing that debt." Hellyer had already established ministerial study groups the year before to examine the status of Canadian defence for the new policy statement and counter calls to reduce the defence budget.[21]

The most significant study group was the Ad Hoc Committee on Defence Policy. Their report, entitled *The Canadian Defence Budget,* concluded that maintenance of the status quo in fiscal year 1964-65 would actually require increasing levels of defence expenditure. They also stated that, if the budget remained constant at $1.6 billion per year, the military would have to reduce personnel in order to fund equipment purchases. The urgency stemmed from the fact that there had been limited procurement of weapon systems since 1957. As the report stated, "the reduction in expenditures on equipment during the last five years has resulted in a backlog of postponed equipment procurement in the order of $500 million. The current rate of spending on equipment is insufficient to make good depreciation and obsolescence. Equipment holdings are therefore becoming increasingly inadequate."[22] The report ended by stating that, if military commitments were to remain at their current levels, procurement of new equipment would be necessary to ensure relevancy in future operations. But as Lieutenant Colonel Ross Fetterly has illustrated, financial limitations created the need for choice: "Defence funding that allocates a significant percentage of resources to capital procurement favours future operations at the expense of current operations." He also explained why this is significant in a liberal democracy:

> Democratic governments usually have a mandate only lasting up to five years. As a result, capital procurement contracts signed in one mandate would only benefit military forces in a subsequent government's mandate. Consequently, in terms of making political mileage, governments tend to favour forces-in-being at the expense of future forces. The result following a period where expenditure has favoured current forces is that the average age of equipment has increased, the cost of maintaining that equipment has multiplied and the backlog of required replacement equipment has increased. This was the situation within which the government found itself in early 1964.[23]

Inflation had caused a weakening of purchasing power in a time when the price of military equipment had skyrocketed. This situation was exacerbated by the fact that a higher percentage of defence money was being spent to keep outdated equipment functioning. Calls for increased cutbacks to defence also continued. At a Cabinet Defence Committee meeting on 21 February 1964, the finance minister suggested "cutting defence expenditure by an additional $500 million and giving the money to the provinces."[24]

The subject of reorganizing the armed forces to save money had also become prominent in the early 1960s. The 1963 *Report of the Royal Commission on Government Organization* contended that budgeting, accounting, supply, construction, and general administration had elements common to all three services and that continued maintenance of the status quo was "uneconomic."[25] Thus began the quest for a single service with a single command structure and a single defence staff. Bill C-90 was passed by Parliament in July 1964, and the position of Chief of the Naval Staff was abolished; the Chief of the Defence Staff position was created for all three services.

This was also a time of increased support for social programs in Canada. The trend of government-funded social programs was the result of a number of factors, including the effects of increased industrialization, a greater desire to reduce poverty, and changing social values. As Fetterly has written, "the growth in the number and cost of social programs and their competition for funding against existing federal programs was a major factor in bringing the significant cost of defence to the attention of citizens and their governments."[26] As the Liberal Party in power brought about a more focused approach to implementing a Canadian welfare state in the 1960s, the RCN had to fight for every dollar and prove that ASW warfare was a worthy investment.

The 1963 *Report of the Ad Hoc Committee on Defence Policy* observed that, since 1959, "to the extent that if there is a national consensus, it is that Canada's defence policies are grossly inefficient, and in need of revision."[27] As a response, the Liberals produced the 1964 Defence White Paper. One of its conclusions was that Canada still had only a modest capability to produce modern weapons systems in the international marketplace and, therefore, must provide continued support for international sharing of development and production of advanced weapons systems, especially with the United States. As Fetterly stated, this was "a pragmatic and realistic approach to the Canadian defence policy conundrum."[28] It was also an acknowledgment that Canada would continue to look to foreign sources for the majority of its military equipment.

Following the 1964 White Paper, the future of ASW operations in Canada looked bright once again. That December the government announced a five-year equipment plan, which pledged to maintain a constant improvement of ASW capabilities. It promised the RCN four DDHs, two operational support ships, one submarine, conversion of the seven *Restigouche* class destroyers to carry ASW rockets, a refit to *Bonaventure,* modernization of detection systems aboard the *Tracker, Argus,* and *Neptune* fixed-wing models, and twelve new *Sea King* helicopters.[29]

By January 1965, *Bonaventure* was sailing with a complement of aircraft that included twelve ASW *Trackers,* one carrier-on-board delivery *Tracker* for general transport, six *Sea Kings,* and one *HO4S* for plane guard.[30] Although landing aboard the large carrier created little difficulty for pilots, by the end of 1965 technical and operational sea trials of the *Sea King* were still under way aboard the less stable DEs. In situations that involved difficult weather, the *Beartrap* was not only necessary to help the helicopter land but also used to straighten and move the aircraft along the rolling deck into its hangar after the blades and tail pylon were folded. The DDH hangars were also widened on both sides to allow for movement around the aircraft. The final tests conducted by VX 10 on the development of the *Sea King* and the DDH concept occurred during 1966, when compatibility trials were carried out from 2 February to 5 March aboard Her Majesty's Canadian Ship (HMCS) *Annapolis.* These tests were to establish further that the DDH could support and maintain helicopter operations both day and night. Although there were still some recommendations and improvements, the trials were an overwhelming success and cleared the way for future operations. From April to June 1966, the *Annapolis* participated in the NATO Matchmaker Squadron, which proved the new capabilities of the Canadian DDH ships.[31] The DDH trials that centred on the *Beartrap* took four years instead of the originally expected four months. But as Cafferky correctly concluded, "certainly, by the late 1960s, the RCN was recognized as the leader in the field of shipborne helicopter operations."[32] The wedding of a modern ASW helicopter to a destroyer/escort-size vessel created a weapons platform that dramatically changed naval warfare.

While the RCN was approaching operational status of the DDHs, there were already plans to deal with developing deficiencies. One area of concern was icing of the rotor blades in bad weather, and work was already being undertaken to solve it. The AQS-10 sonar was also being phased out of service in favour of the AQS-13. In fact, according to Lieutenant Commander R.L. Rogers, when speaking to the Royal Netherlands Navy about Canadian ASW on 7 September 1966, "consideration is now being given to a mid-life modernization programme for this aircraft."[33] Rogers revealed that, though *Sea Kings* had been flying from the carrier *Bonaventure* since 1965, they were not yet fully operational from the DEs. He also stated that the RCN was already looking to update the aircraft with anti-collision radar, magnetic anomaly detection (MAD), Jezebel acoustic ASW sensors to identify submarine class signatures, a stable table navigation system, a navigational computer, and a complete anti-icing capability. ASW technology was moving quickly, and the RCN was determined to stay current.

By 1967, foreign interest in the Canadian *Beartrap* included the United States, Japan, France, and Germany, and the design had become one of Canada's most promising defence exports. The first clearance for service use (CSU) of an operational ship haul-down system was given in November 1966 to the *Annapolis* class HMCS *Nipigon.* The *St. Laurent* vessels had also been completed as helicopter carriers by that time and incorporated all the design changes that the previous helicopter trials had revealed to be necessary. *Annapolis* was subsequently given its *Beartrap* CSU in April 1967, and the rest of the ships followed, with *Margaree* being the last to receive one in November 1968.[34] As a result of the successes of the RCN tests, Germany was using the ship sets to accommodate the Bell *UH-1D* helicopter for search and rescue. Demonstrations to senior American naval officials and military attachés were also carried out aboard HMCS *Fraser* in the Washington Navy Yard on 4, 5, and 6 October 1967.[35] The United States also tested the system aboard the United States Ship (USS) *Catskill,* a mine countermeasures support ship, in 1968.[36] The Canadian sea trials yielded another patent regarding the DE helicopter carriers; Peter Charlton of VX 10 filed an application on 23 August 1965 for rights to the Horizon Indicators for Assisting Helicopter Landings on Ships that had been developed to help pilots keep level during night landings. The patent was granted in Canada on 18 January 1972.[37]

On 1 December 1967, a *Sea King* flew into the sea while on instruments at low altitude. As Stuart Soward explained, "hitting the water at 100 knots is like running into a solid wall."[38] Both pilots survived, one with a broken collar bone, after being thrown through the windshield, but both air crew in the rear of the aircraft were likely badly hurt on impact and sank with the helicopter. This was the first fatal *Sea King* accident. The dangers of flying at sea clearly remained. The *Sea Kings,* however, were still a quantum leap forward in all-weather capability, and this advantage mitigated the risks. In May 1968, for example, HS50 was conducting ASW exercises aboard the *Bonaventure* when some bad weather hit.[39] During periods of zero visibility, the helicopters continued to deliver the ASW screen for the ship and relied completely on their integrated flight control system and the positive radar vector control of *Bonaventure.* The aircraft were able to approach the ship virtually blind until they could detect its wake.

The last of the forty-one aircraft was delivered to the navy on 3 May 1969; by then, it was clear that the *Sea King* would carry out most of the navy's helicopter duties at sea.[40] The Piasecki *HUP-3* retired from service in 1964, the Bell *HTL* helicopters followed in April 1967, and the venerable *HO4S* was retired in 1969.[41] Air crew and ground crew technicians were organized as helicopter air detachments (HELAIRDETS); once the teams were operational, they were assigned to a particular ship. The primary operational deployment of the helicopter was controlled by the captain of the ship, with flight safety and operational flying limits given to the detachment commander, with four-hour missions being the standard.[42] Although several helicopters were assigned to the *Bonaventure* HELAIRDET, one helicopter was assigned to each operational DDH, and flying was conducted by two air crews supported by one maintenance crew. The Base Technical Branch at Shearwater carried out maintenance, and the Helicopter Utility Squadron

21 (HU 21) was responsible for air training. As retired *Sea King* pilot and former Commanding Officer (CO) of 423 Helicopter Squadron Colonel John Orr has declared, "the combination of carrier and DDH platforms provided a logical progression for both aircrew and maintenance personnel with the more experienced personnel serving in the DDHs. As the number of the Navy's air capable ships increased, the manning of HS50 expanded accordingly. By the early 1970s, HS50 was the largest air squadron in the Commonwealth."[43]

The British Royal Navy (RN) had also procured the Sikorsky *Sea King*. The Admiralty was not satisfied with British ASW capabilities in the mid-1960s, and their single-engine *Wessex HAS.3* helicopter was only useful for tracking submarines; it did not have the ability to lift and deliver missiles or depth charges to engage them. Westland Helicopters had acquired a licence from Sikorsky to allow it to develop an anti-submarine helicopter for the RN based on the twin-turbine engine HSS-2 *Sea King*. With the introduction of the American unified aircraft designation system in 1962, the name of the helicopter was changed to the *H-3*. By October 1966, a single *SH-3D* prototype had arrived in the United Kingdom for trials. The *SH-3D* was an upgraded version of the *Sea King* being used in the USN by that time. The basic model procured for the RCN was a *SH-3A*. These models were mostly used for utility missions in the USN as the newer *D* model had an additional fuel tank for greater endurance in ASW operations.[44]

The British then began to modify the aircraft to suit their needs. This version was equipped with British avionics, which included the proven Ekco radar/Type 195 sonar combination from the *Wessex HAS.3*. Westland also replaced the American engines with twin Gnome H1400s and added a fully computerized control system and tactical display that was superior to that of the USN and the RCN.[45] The RN was extremely efficient at incorporating the avionics package into the new *Sea King* airframe. It developed, tried, and tested it in the *Wessex* model while the *Sea King* airframes were being built. This judicious course of action alleviated the necessity of starting developmental work after introduction of the new helicopter. Colonel (retired) Lee Myrhaugen was the Canadian Exchange Officer at Royal Naval Air Station, Culdrose, at the time of the introduction of the RN *Sea King*. He explained that this procurement approach "was highly effective in allowing the RN to bring the Sea Kings into operational service in record time."[46] The first British HAS Mk1 *Sea King* was commissioned in August 1969, the year that the last Canadian *Sea King* was delivered. In all, there were twenty-five different versions built for the Fleet Air Arm of the RN and eight other armed forces.[47]

The issue of maintenance was of paramount importance for the *Sea Kings* in all naval services. Helicopters are complicated technological platforms and must be well serviced to avoid putting the air crew in danger. In Canada, most major maintenance was initially to be carried out at Shearwater, and room for spare parts was made aboard the DDHs; even spare engines and rotor blades were kept on hand. As Peter Charlton and Michael Whitby have written, "the end result was that the detachments aboard ship were able to do almost all of the work required to keep the aircraft serviceable and usually with a higher availability than was achieved ashore. This is of course typical of such detached operations where there are few distractions and long hours of work become the norm."[48] When more space was required for heavy maintenance, *Sea King* detachments were also stationed aboard the supply ship *Provider* and later the Auxiliary Oil Replenishment (AOR) ships *Protecteur* and *Preserver* after they were commissioned in 1969 and 1970 respectively. Their hangars held up to three *Sea Kings*.[49] The final trials aboard the smaller DDHs for night and heavy weather operations were carried out on HMCS *Assiniboine* in 1970, and full CSU followed shortly after. Herb Harzan, a project test pilot with the VX 10 team at the time, stated that "it was at that time that the Canadian Forces were so far ahead in operational capability of operating helicopters from destroyer size ships in comparison with all other Navies."[50]

On 8 May 1967, Bill C-243, the Canadian Forces Reorganization Act, was passed. The naval air arm legally ceased to exist when the Canadian Unification Act was proclaimed law on 1 February 1968.[51] The navy itself was no longer a separate entity from the army and the air force but part of the new Canadian Armed Forces (CAF). One of the key topics within the new CAF was the future of carrier-based operations. One of the government promises that stemmed from the 1964 White Paper included refitting HMCS *Bonaventure*; however, the refit actually led to criticism of the platform because of massive overspending. The overhaul was carried out between April 1966 and August 1967 and was originally to cost $8 million. The twenty-three-year-old hull, main plumbing, and ring-main water systems, however, were revealed to be in desperate need of repair or replacement. The cost for these major projects was not anticipated in the original estimates, and there was an attempt to disguise them under other repairs. The end cost of the project was over $17 million – more than double the original estimate – and the ship was out of service for nearly eighteen months. As Soward has declared, "the most humiliating aspect of the debacle was not that there were $9 million spent in excess of the estimates on necessary repairs; the real tragedy was that no one attempted to justify the cost to the public. Bearing in mind the complete confusion at NDHQ [National Defence Headquarters], it is likely that no one really felt responsible at this point."[52] The refits were subsequently investigated by the Parliamentary Public Accounts Committee and targeted by many as a waste of capital.[53] *Bonaventure* was scheduled for withdrawal soon after. Despite being in top condition after its modernization, the carrier was retired in 1969.[54]

It has been postulated that, as unification began to undermine the control of naval aviation by former RCN personnel, a trade-off agreement was made by the former RCN and the Royal Canadian Air Force (RCAF) senior officers regarding naval aviation. The tacit agreement was that all fixed-wing aircraft would be controlled by the former RCAF personnel in Maritime Command, and the maritime commander would receive the support of the former RCAF in the conversion of the DEs to ASW helicopter carriers. According to one author, the maritime commander subsequently discussed the future of Canadian naval aviation with senior aviators in the context of helicopter carriers. Soward concluded that this was "a development which would be disastrous for naval aviation."[55] Although helicopters were obviously involved in airborne operations at sea, they were not usually discussed within the scope of the term "naval aviation." As Colonel (retired) John Orr has put forth, "rotary wing aviation was considered to be a sideshow compared to fixed wing aviation by senior Naval Aviators who had all cut their teeth on fixed wing operations."[56] All the same, the aircraft carrier was seen by the government as an expendable platform in the age of the DDH. As Michael Hennessy has proclaimed, "facing retrenchment and financial austerity, the RCN planned for missions confined largely to ASW in the Western Atlantic."[57] And the new helicopter-carrying DE would assume this role from *Bonaventure*. It was understood as early as 1953 that helicopters could be used to expand the capability of the navy and make up for the lack of new ships.[58] Although the platform did change, the aviation capability did diminish with the retirement of fighter planes, and the term "naval aviation" was no longer used, the sphere of air operations being conducted at sea did not completely vanish. The decommissioning of the carrier, however, directly affected the use of helicopters because it could no longer be used as a way of training and easing pilots into the more difficult area of operations aboard a DE. The carrier also transported the maintenance equipment that could not fit aboard these smaller vessels.[59]

After *Bonaventure* was retired, ASW helicopter operations underwent a permanent change. Orr has explained that, "for major exercises, ships normally sailed as part of the *Bonaventure* Task Group and flight operations were controlled by the carrier. When the carrier was not deployed, the DDH's sailed as part of a Destroyer Squadron with overall control of the embarked helicopter's flying progamme exercised by the Officer in Tactical Command. Only on infrequent occasions did ships sail independently."[60] After 1969, these Canadian destroyer squadrons (CANDESRONS) soon became the norm. Flight operations in the DDHs often required the helicopter to fly for a twelve-hour block with a refuelling and crew change after a three-hour mission. Pilots honed their skills through an annual cycle of exercises that included individual ship and helicopter combat readiness training in the Puerto Rico Operating Areas as well as national, North American, and NATO exercises scheduled throughout the year.[61]

As helicopter-based operations became the focus for ASW operations at sea, the hours logged in the aircraft increased, and, as an obvious corollary, they required more service and placed considerable strain on the maintenance personnel. The entire fleet was actually grounded in 1969 because of a gearbox defect.[62] The 1970 Maple Spring exercise in Puerto Rico also saw all five embarked *Sea Kings* become unserviceable, and they were fixed just in time for their exercise commitments.[63] One significant transformation caused by unification of the forces was that centralized maintenance was introduced to Shearwater to bring the base into conformity with Canadian Forces air maintenance practices. Aircraft and maintenance personnel were now under the supervision of the Base Aircraft Maintenance and Engineering Officer (BAMEO). Operational training for the *Sea King* was also revamped with the formal establishment of VT 406 (later HT 406) in 1972 as a full Operational Training Unit responsible for both flying and technical training.[64] There are differing opinions on the relationship of the "light blue" and the "dark blue" that resulted from such integrations of maintenance crews. Dudley Allan, a qualified marine and air engineer at the time, declared that "the theme comes up time and again in talking to ex-Naval maintenance trades on their experiences post unification with the RCAF counterparts; how much better the Navy training was. Well, it had to be – we had to do more with less and serve at sea miles away from base support. That experience stood us all in good stead."[65] Indeed, it is clear that integration of the services made maintenance of the helicopters more difficult as the technicians had to be cross-trained. As Gordon Moyer, a base engineer at Shearwater in the 1970s, asserted, "this made things more difficult, as the Air Force guys knew little about helicopters."[66] Regardless of which servicemen were more skilled, it was clear that the *Sea King* had always been a high-maintenance aircraft, and it was a demanding task to keep them fully functional.

After Trudeau and the Liberals came to power in 1968, a Management Review Group (MRG) was appointed by the new Minister of National Defence, Donald Macdonald, to examine all aspects of the management of the Department of National Defence (DND). Their final report was entitled *The Management of Defence in Canada* and was submitted in July 1972. It stated that

> Effective coordinative management is lacking despite the obvious need and, in consequence, there is no focal point of accountability for performance. Since clear Departmental policies in matters related to defence procurement are not provided, political and economic considerations do not always guide effectively the specific procurement programs. And it is because of failure

> to adequately assess the financial and technological risks inherently faced in procurement programs that there is a corresponding failure to accommodate those risks in either the planning or administration of the program as events unfold.[67]

The MRG subsequently recommended that all research, engineering, and procurement be consolidated under one Assistant Deputy Minister (Materiel) (ADM Mat). Before the report was published that year, it was announced by a new Minister of National Defence, Edgar Benson, that the military and civilian elements of the DND and Canadian Forces Headquarters would be integrated into a new NDHQ. The new position of ADM (Mat) was created within the new organizational structure. Although this position was designed to facilitate effective procurement, some officers also thought that it was part of the civilianization of the DND and that it only added to the bureaucracy. Colonel (retired) John Orr has asserted that, "when I showed up in NDHQ in the mid-seventies the procurement process became more and more muddled as the programme-management aspects fell increasingly to the ADM (Mat) organization."[68]

Adding to the procurement problems of weak project management, risk assessment, accountability, and an overall lack of policy outlined by the Management Review Group, the money for new equipment disappeared in the early 1970s. Canada had entered a period of prolonged peace, and Trudeau decided that the Canadian defence policy had been based solely on NATO policy. His government subsequently decided that the European members of NATO could protect themselves and that Canadian defence commitments and the budget could be reduced. From 1964-65 to 1972-73, the proportion of defence expenditures devoted to equipment procurement went from 14.8 percent to 5.9 percent; 25 percent was often used as the minimum needed to maintain military forces in Western industrialized states.[69]

The only bright spot for procurement in the early 1970s was the four new *Tribal DHH 280 Iroquois* class gas turbine destroyers that arrived in Halifax in 1972. The ships had been promised by the government in 1964. The savings that resulted from *Bonaventure*'s retirement made it possible for the navy to revitalize part of its surface fleet with the procurement of the new destroyers. Commissioning of HMCS *Iroquois,* the first to enter service, took place on 29 July. The ships were expressly designed and produced in Canada for integrated helicopter ASW operations.[70] But as with the other shipbuilding programs, the one for the *280* was characterized by inept project management and unchecked cost overruns. The quest to reorganize and integrate the Canadian Forces created a management vacuum at the DND, and it was unclear who had control of the project. The original estimate for the program was $142 million. By the time the ships were complete, the project was revealed to have cost $252 million. Trudeau was even forced to create a special Treasury Board Committee to investigate. According to Marc Milner, "by 1970, the embarrassment was so great that the whole program – despite the fact that the hulls were rising – was nearly cancelled. Only the fact that 75 percent of the program funds had already been spent, together with the announcement of a firm final price of $252 million, saved the ships."[71] Despite this situation, the ships were extremely capable once they entered service, and there has been much praise of their abilities.[72] These ships were capable of embarking two helicopters, and the three air crews could fly both aircraft for a combined total of eighteen hours per day in a twelve-hour block. Two maintenance crews were provided, which permitted twenty-four-hour servicing and first-line maintenance. "The two aircraft and expanded maintenance capability meant that the *Iroquois* class ships could generate and sustain a considerably higher rate of air effort than the single aircraft HELAIRDETs of the *Improved St. Laurent* and *Annapolis* classes."[73] But as one task force report concluded, "in 1974-75 the major equipments [sic] in service were at or near the end of effective service and life extensions had become costly ... The destroyer fleet, other than the four new DDH 280s, had limited life and were obsolete."[74]

With the entry of the *Iroquois* class, there came an increased requirement for additional ASW helicopters, more air crew, and more maintenance personnel. By this time, HS50 had become too large to function effectively as a single unit. Consequently, it was decided that the unit must be divided into two squadrons and expanded overall. HS50 was thus disbanded, and in its place 423 and 443 were resurrected as Helicopter Anti-Submarine Squadrons. As the "History of 423 Squadron" put it, "thus, on 3 September 1974, 423 Squadron, a World War Two anti-submarine squadron was once again in the business of hunting for submarines."[75] Lieutenant Colonel Larry A. Ashley took command of HS 423 and with Executive Officer Major Jav Stevenson began the task of building an administrative and training organization.

423 Squadron was to support CANDESRON ONE, provide five HELAIRDETs for HMCS *Iroquois, Saguenay, Annapolis,* and *Fraser,* and establish a headquarters based in Shearwater. 443 Squadron was to support CANDESRON FIVE. Part of the unit's mandate was to be "responsible for operations in support of the Navy's responsibilities and requirements for surface and sub-surface surveillance, maritime warfare, cooperation with other commands, forces and agencies, air/sea rescue and other missions that may be assigned."

In September 1975, exactly a year after the split of HS50, another significant change took place with the establishment of Air Command. Full command of the *Sea King* fleet was then transferred from Maritime Command to Air Command. Operational command remained on the coast in the hands of the Commander, Maritime Air Group. Once again it was the increasing loss of control that galled naval aviators. Vice Admiral J.C. O'Brien lamented that "the whole Goddamn thing has been given back to them [Air Command] ... We don't even own the goddam Sea Kings – and they are part of the ship's weapons system! I don't understand it. There is no naval background to the aviators; they're just fliers and don't really care what goes on."[76]

The Task Force on Review of Unification of the Canadian Forces later discovered that many ex-navy felt that the "airmen" did not participate in the ship's normal routine as they had not been trained as sailors, that their adjustment to naval life was difficult, and that they hurt the cohesiveness of the ship's company.[77] The first Commander of Air Command, however, Lieutenant General Bill Carr, recalled his first visit to Shearwater with fondness. After he flew a night ASW mission with one of the junior crew commanders, he was effusive:

> I'm still impressed with the "routine" demonstration I was put through over the rainy, pitch black, and rough fluorescent Atlantic at fifty feet well out to sea off the coast of Nova Scotia! During my flying career I had been exposed to many different kinds of operations, but none had impressed me more than this professionalism which [the] Sea King operation demanded and regularly demonstrated.[78]

But regardless of this praise, there were still antagonisms between the services within the system of unification. As Colonel (retired) Laurence McWha explained,

> Following the formation of Air Command in 1975, it was permissible to refer to ourselves as "Shipborne Aviators" and to our trade as "Shipborne Aviation" but in the higher circles of the new Air Command and particularly amongst the old bulldogs of the ex-RCAF, our use of these terms to distinguish ourselves from the rest of the "Air Force" was barely tolerated. It took many years of Unification dust-settling before these terms of self-description became an accepted part of the Air Command lexicon.[79]

There had always been aggressive statements by former RCN aviators decrying the "perils of unification" or how "unification reared its ugly head" and how they detested the new terminologies and the new levels of organization. But while the members of the former RCN had to adjust to an incursion by former RCAF personnel, and there were many incidents of interservice rivalry, there was also a level of understanding between former RCN and RCAF personnel regarding naval helicopter operations. As the "History of 423 Squadron" has claimed, "the helicopters were a valuable extension of the ship's combat sensors (finding the enemy), and attack capabilities (destroying the enemy). The RCN Air Arm had proved these capabilities, and as 'their' assets were integrated into the Air Force, they were wary of what the future might hold. A new era of Navy-Air Force cooperation, however, was to begin."[80] Perhaps it can best be described as an unwanted situation that neither service fully accepted but had to learn to function within. There were bigger problems ahead for what remained of naval air operations; the very aircraft relied on at sea were in need of modernization or replacement.

Although the *Sea Kings* were arriving in Canada in the early 1960s, the trials aboard the new *St. Laurent* class vessels to test the new *Beartrap* haul-down system took far longer than expected. It took time for VX 10 to work out the many prototype flaws that existed and how best to decide on the final product that would allow the new aircraft to function in tandem with Canada's destroyers. The Canadian *Beartrap* was a new technology, and additional costs and faults in its early design and production were to be expected. Instead of four months, it took four years. Moreover, it was revealed that there was an overall lack of program coordination in 1963, and the RCN could not acquire sufficient American training facilities. The delays and cost overruns were the common result of a foreign equipment purchase. But procurement of the *Sea King* was a success, and the technological advance of the *Beartrap* system created a weapons platform that changed how Canadian naval warfare was conducted. During the 1960s, the Canadian government decided to end carrier operations and concentrate on helicopter-carrying ships instead. From that point onward, the *Sea King* helicopter was placed in a central role in maritime defence. By the end of the decade, the navy was recognized as the international leader in the field of shipborne helicopter operations, and the other major navies of the world were forced to take a serious look at Canadian ingenuity. As fixed-wing ASW aircraft were phased out in Canada, the operational tempo of the *Sea King* increased. A major disadvantage of this new focus on maritime helicopter operations, however, was that, since the aircraft were logging more hours in the air, they logically required increased levels of maintenance and service at sea.

Crewmembers from HMCS *Winnipeg*'s helicopter detachment prepare a CH-124 *Sea King* for flight operations in the Gulf of Oman on 31 July 2005. The Canadian frigate is part of Operation ALTAIR, Canada's maritime contribution to the US-led coalition campaign against terrorism mission known as Operation Enduring Freedom. HMCS *Winnipeg*, operating with the 5th Fleet of the United States Navy, is deployed for a period of six months to the Persian Gulf (Arabian Gulf) region. With her crew of 240 officers and sailors and her CH-124 *Sea King* helicopter detachment, HMCS *Winnipeg* is conducting surveillance patrols and maritime interdiction operations in order to control sea-based activity in the region.

Source: CF photo by Sgt. Frank Hudec. *Courtesy of National Defence. Reproduced with the permission of the Minister of Public Works and Government Services, 2009*

5
The New Shipborne Aircraft Project: A Commitment to Replace the Fleet

> The Sea King Helicopters now in service are already at the end of their useful life and a process to select a new shipborne aircraft, to be produced in Canada, is currently underway.
>
> – 1987 Defence White Paper

A major *Sea King* overhaul program began in 1975. It was called the *Sea King* Improvement Program (SKIP). It provided updated engines and an improved transmission, along with the installation of a weather avoidance radar.[1] SKIP also included the installation of a crash-worthy fuel system, rotor blade de-icing, strengthened crew seats, an improved Tactical Air Navigation/Distance Measuring Equipment (TACAN/DME) readout, and an AN/APS-503 radar.[2] From 1975 to 1977, the program also added sonobuoy and marine marker chutes, dipping sonar improvements, and the ability to hover-drop torpedoes. Sonobuoys were four-foot-long cylindrical devices that floated vertically in the water and used small radios to transmit sound recordings up to an aircraft. After they transmitted for a few hours, they simply sank to the bottom of the sea. At the conclusion of the upgrades, the official designation of the *Sea King* was the *CH-124A*.

Some of these improvements, however, were ineffective, and other problems were still ignored. The helicopter, for example, still lacked an appropriate acoustic processor for the level of warfare being practised at the time. Modern helicopters, for example, were using new acoustic processing systems that categorized underwater sounds against an extensive internal library of sonar "signatures." Each vessel had a distinct sound, like a fingerprint, which could be used to determine friend or foe.[3] The simple fact that the *Sea King* was designed with 1950s technology made the aircraft increasingly ineffective against modern submarine warfare. Colonel (retired) John Cody has since called the upgrades the "get well programs of the 70s and 80s."[4]

As SKIP progressed throughout the late 1970s, it became clear to naval officials that continued refurbishing of the aircraft would be very expensive. Moreover, the relative lack of speed and endurance, as well as the lack of a digital acoustic processor, could not be solved with any level of upgrade or retrofit.[5] The first discussions on the replacement of the *Sea King* had already begun in 1975, and the first Statement of Requirement (SOR) for a replacement was forwarded from Maritime Command to National Defence Headquarters (NDHQ) that year.[6] The entire fleet clearly did not need immediate replacement as the last *Sea King* had entered service only in 1969; this early planning could be considered the result of overzealous staff work by the navy. As Stephen Priestley has written,

> Although the Air Force began to discuss the eventual replacement of the Sea King in 1975, it was not a major priority even by 1978 as other procurement plans had priority – at the time, new fighters and long-range patrol aircraft ... And, perhaps, Air Force planners believed that they had time to spare for replacing those aging helicopters. Whatever the reason, the proposed *Sea King* replacement remained a distinct project and was given a rather low priority. After all, hadn't the entire CH-124 fleet just been through a thorough structural rebuild and been re-fitted with the latest in Canadian ASW [anti-submarine warfare] equipment. What's the rush?[7]

But the *Sea Kings* had been designed in the 1950s, so it was not strange that planners were beginning to discuss future possibilities in 1975. Moreover, considerations for equipment replacement must always begin early as the procurement process for major acquisitions had already revealed itself to be exhaustive with the original *Sea King* purchase.

There were also other projects ahead of plans to replace the *Sea King*. The three services of the Canadian military all required equipment upgrades to keep pace with developments in their respective areas of operation. As a result, each had to take its turn in regard to major acquisitions and compete with each other for scarce resources. By 1975, the Trudeau government launched a re-equipment program as part of the 1974-75 Defence Structure Review, and the first major project was for the CP-140 *Aurora* long-range patrol aircraft (LRPA). It was approved in July 1976, and it "was the first to move beyond the traditional 'Canadian Content' provisions of previous offshore defence procurement programs, and to stipulate in elaborate contractual language (including clearly specified financial penalties for non-compliance) the need for the foreign prime contractor (Lockheed Aircraft Corporation) to attempt to achieve a wide variety of economic objectives over the lifetime of the program."[8] The government employed a funded Contract Definition Phase to increase the competition between potential contractors and the level of "offsets," or reciprocal investments in Canada, for the project. This policy invited a myriad of other departments into the procurement process to ensure that Industrial and Regional Benefits, later known as IRBs, were being addressed within the military's large acquisition programs.

This industrial policy was also extended to the procurement of the C-1 *Leopard* main battle tank built by Krause-Maffei in West Germany. The *Leopard* was to replace the British-built *Centurion* tank and was approved on 30 September 1976. The obvious lack of a tank industry in Canada, however, combined with the relatively low cost of the project gave the government little leverage to negotiate with the German company. The contract was for $187 million compared to the $697 million spent on the *Aurora*. The final offsets amounted to only 40 percent of the contract price, whereas they were 96 percent for the *Aurora* program. Moreover, Krauss-Maffei did not spread its offset purchases widely across Canada. Notwithstanding the comparative level of success of the two contracts, both made explicit the Canadian government's desire to tie non-military economic factors to defence purchasing, and IRBs became inextricably tied to defence procurement in Canada.

The *Sea King* Replacement Program (SKR) was established as part of the Ship Replacement Program, announced in December 1977, to enhance the Canadian frigate's ability to combat modern submarines at extended ranges. ASW at the time was all about speed, range, and new sonar systems. It was hoped that new ships and helicopters could detect submarines beyond the range of their missiles. Marc Milner has noted that

> Therefore a big helicopter, faster, and with greater range than the Sea King, was needed to localize contacts at extreme range, and it formed the centerpiece of the new frigate's weapons system. Canadians also wanted an aircraft that had airborne early warning and perhaps provide mid-course guidance to missile systems. It was understood that by the time the first frigates entered service in 1985, the Sea Kings would be twenty years old and due for replacement. In 1977, therefore, planners were already thinking about a new helicopter.[9]

Indeed, Brigadier General Roy Sturgess, Commander of the Maritime Air Group in 1978, claimed that the *Sea Kings* would be replaced by 1985.[10] The SKR, however, was unbundled from what became the Canadian Patrol Frigate (CPF) project in 1978 due to its complexity and cost and proceeded as an independent project.[11]

In July 1978, the SKR was registered in the Defence Capital Program. On 20 August 1979, the SOR was formally approved, and the program planning proposal (PPP) was accepted in March 1980. From that point, the SKR was renamed the New Shipborne Aircraft (NSA) project on 1 January 1981 and was reregistered into the Defence Capital Program under that title.[12] Once again, however, this did not signal a dedicated effort to carry through on a new procurement. Gordon Moyer was part of the inchoate project management team in 1983-84 and has since declared that "it was like pissing in the wind to try and get it going ... By this time, everything was breaking down. It was always a labour intensive aircraft. With all that banging around, it was difficult to know what the problem was."[13] Indeed, the *Sea King* sustained a great deal of turbulence during operations and required not only upgrades to its ASW equipment to keep it operational but general airframe overhauls as well. By the late 1970s, the mission suite was no longer effective against modern Soviet submarines.[14]

Although the first *Sea King* did not enter service in the British Royal Navy (RN) until 1969, the helicopter fleet was being upgraded at the same time as the Canadian aircraft. The *Mk 2* version had been fitted with new engines, and a six-blade tail rotor replaced the old five-blade type. The engine intakes were also shielded, and fluid de-icing strips were added for better protection. These modifications were complete by June 1976. Further upgrades were done in 1979, and the *Mk 4* version included a revised avionic fit with the Decca Doppler 71 and a new Tactical Air Navigation System. The next year sonobuoy, sonar, and radar improvements were added to the *Mk 5* to complement dunking sonar, and the modernized *Sea Kings* were fully functional in time for the Falklands Conflict of 1982. A modern airborne early warning (AEW) "searchwater" radar radome was also added to some of the *Sea Kings* that year.[15]

The sheer size of the United States Navy (USN) and the scope of American military operations meant that its *H-3 Sea Kings* were constantly being modified and upgraded. For example, in December 1965, aerial refuelling tests were conducted on the *CH-3C.* A refuelling boom was subsequently added to the centre of the nose of the aircraft for the production of the *HH-3E.* Newer versions were also consistently purchased to replace models considered obsolete in the field. For example, as the Vietnam War escalated and inshore bombing missions were carried out from the Gulf of Tonkin by US naval air forces, the need arose for an armoured combat Search and Rescue (SAR) version. Called the *HH-3A,* it operated during the 1970s and 1980s. The primary USN version of the *Sea King* for ASW during 1971 was the *SH-3G.* Although that version had just come online, further improvements had been made by the end of the year. The conversion included an AQS-81 towed magnetic anomaly detection (MAD) system, electronic surveillance equipment to detect incoming missiles, and an external chaff/flare dispenser for airborne protection against ground/sea-to-air missiles. This version was called the *SH-3H,* and after it proved highly successful at sea the earlier models of

SH-3A and *SH-3Ds* were also converted to that standard. This conversion was completed by 1976.[16] Canadian Colonel Laurence McWha has observed that "the UK and the US forces continuously upgraded their Sea King fleets to replace older systems with newer ones for reasons of improved capability and supportability. Canada, on the other hand, did not spend a cent unless a component became obsolete or a safety requirement for new equipment could be argued."[17]

The issue of extensive maintenance involved in the operation of the Canadian *Sea Kings* was explained by John McDermott, former Commander of 423 Helicopter Squadron. He also revealed that there was a lack of availability of aircraft in the 1970s:

> It may surprise current Eagles to know that even in the mid 1970s we had aircraft availability problems. Our problem was that the Sea King was undergoing the second part of its mid-life update. The NDHQ plan was that 7 aircraft at a time would be removed from service and sent to [the] contractor for an upgrade program that would last approximately three months per aircraft. In fact, it took the contractor, on average, six months to complete the work on each aircraft. That would have been okay if we were only short 7 aircraft at a time. Unfortunately, whoever had negotiated the contract specified that aircraft be delivered to the contractor on specific dates, whether work on the preceding aircraft had been completed or not. To avoid penalty payments, we delivered aircraft as required, resulting in far more than 7 aircraft out of the system at once ... That is not to blame the maintainers or their organization, it just reflects that for its entire life, the old Sea King has been difficult and unpredictable.[18]

With fewer operational aircraft, air crews and equipment were shifted between detachments as ships' programs changed. Short-notice changes were the norm as 423 Squadron scrambled to meet its tasks. This was increasingly a problem by 1977 as Canada had claimed a two-hundred mile Exclusive Economic Zone. More surveillance to protect sovereignty and Canadian coastal interests was needed, especially regarding the nation's fisheries. Maritime Command and the surveillance capability of the helicopter-carrying destroyers (DDHs), therefore, became invaluable in supporting Canadian foreign policy objectives related to the law of the sea.[19]

Through the added frequency of sovereignty patrols by ship and aircraft, the helicopter was clearly established as an important extension of Canadian destroyer squadrons in that role. Increased activity meant increased training. In 1976, it was revealed that, during preparations for the spring exercises, "the most valuable personnel asset at this time was someone who knew where to acquire the multitude of equipment and spares needed to keep aircraft flying during a busy exercise. Begging, borrowing and 'otherwise obtaining' became the order of the day for the detachment scroungers."[20]

By the late 1970s, there was a history of in-flight engine failures aboard the *Sea King*. Glenn Cook took command of 423 Squadron on 12 July 1979, and on his first flight to Ottawa one of the two engines failed, and he was forced to make an emergency landing in Quebec City.[21] Cook recalled: "I knew the Sea King could not maintain 5,000 feet of altitude on a single engine and immediately started to dump fuel overboard in order to reduce weight." Although he landed safely, he revealed that it had been the eighth or ninth *Sea King* engine failure that he had encountered in his career. Lieutenant Colonel Ron Holden was also aboard and "expressed little further interest in flying in the *Sea King* and made arrangements to fly to Ottawa via Air Canada." After he assumed command, Cook also discovered that the antiquated radar altimeter warning system for low-level flying at sea had a mean time between failure (MTBF) of 10.4 hours; the standard was 1,000 hours. Restrictions on night dipping followed for the next eight months as crews waited for a replacement.

One promising development was the establishment of the Helicopter Operational Test and Evaluation Flight (later Facility) (HOTEF) at Shearwater in 1980. In addition to routine acceptance and trials, this unit became involved in a number of important developmental evaluations associated with new equipment for the later replacement projects. As Peter Charlton, the former technical officer of VX 10, explained,

> Its mandate was to conduct evaluations of prototype avionics and systems for the shipborne helicopter community of Maritime Air Group (MAG) ... With a staff of some six Pilots, six Navigators, two Airborne Sensor Operators, an AERE [Aerospace Engineer] Officer and a number of Technicians, HOTEF [had] developed a powerful capability to continue the work begun by VX10 into the era of the ... replacement of the Sea King.[22]

Problems with the actual aircraft began to intensify in August 1982, and they dictated the cancellation of squadron involvement with maritime deployments for the rest of the year. There was a problem with the main transmission support brackets, and the fleet was grounded most of the time until December. As the 423 squadron history proclaimed, "proficiency and operational readiness suffered from the lack of flying hours, but spirits remained high. When *Sea Kings* returned to service in the new year with replaced or retooled transmission mounting brackets, 423 quickly regained its operational skills."[23] Although the ability of Canadian helicopter pilots was not in question, the performance of their aircraft was. In 1985, Martin

Shadwick of York University stated that "it may be, in fact, that there is no practical alternative to the earliest possible replacement of the Sea King fleet."[24] Thomas Lynch agreed:

> Canada, with a defined role that still emphasizes an antisubmarine warfare role, has found herself in an embarrassing position with the aging *Sea King*. Once on the cutting edge of technology by placing a relatively heavy ASW helicopter aboard a small warship such as a frigate, the Canadian Navy has seen its leadership eroded and surpassed by successive generations of helicopters in both the USN and RN.[25]

During the spring and summer of 1985, the decision was formally made by the Department of National Defence (DND) to replace the *Sea King*. On 23 May, the project was recommended for approval by the Program Control Board of the DND; on 10 June, it was approved departmentally by the Defence Management Committee; and finally, in August, the NSA project office opened. Colonel McWha has stated that there was "a de facto project office in place long before that date busily drafting the necessary specifications and project planning/approval documents."[26] He also explained that the official project office was delayed because the procurement process at the DND had become unwieldy and required unremitting approval and analysis at many levels. This dilatory process continued as the SOR that was first accepted for the SKR in 1979 was subject to another review after the new NSA project office was opened.

In April 1986, the DND authorized the issue of a Solicitation of Interest (SOI) package to industry. It outlined what the military was looking for regarding the NSA project and was given to all major aerospace companies believed to be in a position to meet the NSA requirement. A submission to the Treasury Board was made in May. On 5 August, the Canadian government approved the Project Definition Phase of the NSA.[27] After the application of minor amendments and reformatting, the new SOR document was reapproved for the NSA on 1 October, and the official Request for Proposal (RFP) was released to ten companies on 12 November. It was an eleven-volume document that outlined what Canada needed in a maritime helicopter, and only a few of the companies were expected to compete. The increased complications to the procurement process within the DND were obvious; the stated requirements for the first naval helicopter procurement in 1951 had been written on only a couple of pages.[28] The RFP required a helicopter that had a 20,000- to 30,000-pound all-up weight and that would carry a crew of four or five. It had to be multi-engined, be capable of ASW, ship surveillance and targeting, and be capable of secondary roles of SAR, medical evacuation, and vertical replenishment. It had to have modern self-contained navigational abilities and be capable of operating in high-density traffic areas. Moreover, it had to have all-weather functionality and be night operable with increased speed, range, and endurance over the *Sea King*.[29]

The procurement process was to have a "Definition Phase" and an "Implementation Phase." The principal aim of the former was definition of a "fully costed weapon system" that substantiated the number and type of aircraft that best met the NSA SOR. What is significant here is that the NSA office had decided that the procurement would be aimed at the employment of a "Canadian prime contractor," vehicle manufacture or assembly in Canada, system integration in Canada, DND-funded research and development projects, and the establishment of domestic lifetime maintenance.[30] The Implementation Phase would consist of the development, qualification, and production of a fully supported helicopter by 1994. The precedent for the inclusion of a Canadian prime contractor came from the CPF project, which had been awarded in 1983. The CPF project office did not simply want Canadian industry to participate in fulfillment of the contract – it wanted a Canadian company to manage it.[31]

The 1985 Nielsen Task Force Report had made recommendations that the government needed to be much more expansive in its search for benefit packages. The Canadian government subsequently looked to spread the IRBs beyond central Canada, receive higher degrees of technology transfer, increase domestic development and the industrial competitiveness of Canadian firms and their export opportunities, and have life-cycle support contracts established in Canada. It also wanted to apply these requirements more consistently to major purchases. Defence contracts had to have explicit political value and to reflect the nature of the Canadian federal system; the government had to show it was not favouring any particular region. Overall, this policy was seen to increase risk for weapons acquisition, particularly the acceptance of equipment still in development. But it also increased the possibility for more sustainable, high-quality IRBs.[32]

By 1986, IRBs had become permanent policy in Canada, which was quickly transferred to the NSA. Clearly, the government wanted Canada to be thoroughly involved in the procurement of the new helicopter, and any foreign company that bid on the NSA had to make an economic partnership with domestic industry that far surpassed the early IRB examples such as the *Leopard* contract. As was demonstrated with the original *Sea King* acquisition, however, even the domestic assembly of a foreign design was not without challenges. And the NSA project office wanted much more than that. The NSA Project Manager, Colonel David Bennett, stated at the time that "we do have some desire that the prime company itself is Canadian ... It's also a strong desire for us that there be regional distribution of the industrial participation."[33]

In 1987, a Defence White Paper was issued by the Mulroney government. It recognized that the influence of a nation was partly dependent on its investment in collective security and set out to reverse the damage done to the Canadian Forces (CF) through budget cuts since 1971. A large part of the paper focused on the acquisition of new equipment. The White Paper acknowledged that a significant "commitment-capability gap" existed and that it prevented the military from carrying out the government's mandates.[34] It asserted that "failure to provide modern equipment has undercut the credibility of the Canadian Forces ... Moreover, if Canadian men and women ever had to go into combat with the aged equipment they currently possess, lives would be needlessly lost. Decades of neglect must be overcome." It also recognized that defence planning is a long-term process and that the government needed to provide a planning framework wherein the equipment decisions did not lead defence policy.

A new defence investment framework was also established to link it with domestic industry. The White Paper explained that,

> In cooperation with other departments, the Department of National Defence will pursue policies to increase R&D capability and expertise in Canada. The Defence Industry Productivity Program of the Department of Regional Industrial Expansion is an example of such an initiative. Canadian technology will be incorporated into National Defence operations and plans wherever possible in order to help strengthen Canada's defence industrial capacity.[35]

Since one of the goals was also to pursue an aggressive naval modernization program, the Conservatives continued the pursuit of new patrol frigates and repeated the call for new shipborne aircraft. The paper conceded that the *Sea King* helicopters were already at the end of their operational lives. Any modern ASW helicopter was clearly going to be purchased from a foreign source, and as the White Paper maintained, "where major equipment must be procured off-shore, the government will promote teaming arrangements with Canadian industry to foster technology transfer and the creation of an indigenous support base." For this to occur, it was recognized that a concerted effort to "train additional specialized people" for procurement program management was necessary. Douglas Bland has also explained that a primary responsibility of the Assistant Deputy Minister (Materiel) (ADM Mat) as part of this policy was "to facilitate the development of industrial capabilities."[36] The policy that linked defence procurement to domestic industry was to profoundly affect the replacement of the *Sea King*.

The two main companies that responded to the RFP guidelines for the NSA were European Helicopter Industries (EHI) and Aerospatiale. Both submissions were made in February 1987.[37] Aerospatiale offered its SA 332F1 *Super Puma* Mark II. EHI offered the *EH-101*. In 1977, the Royal Navy had placed a requirement for an ASW helicopter to the Ministry of Defence (MoD) to replace its *Sea Kings,* which had been built under licence by the British company Westland. Westland Helicopters submitted its *WG.34* project to the RN, and it was then approved for development. At the same time, the Italian navy needed to replace its version of the *Sea King,* which had been built under licence by the Italian company Agusta. Discussions then took place between the two companies during 1980, and they formed the company EHI. This partnership was specifically formed to produce the *EH-101*. In 1984, the British and Italian governments signed the agreement, which provided joint funding for the development stage straight through to the production stage. The *EH-101,* therefore, was being designed expressly to replace the British and Italian *Sea Kings,* and prototypes were under way by 1987.[38]

Both EHI and Aerospatiale were required to comply with the Canadian policy of linking foreign equipment purchases to domestic industry, as outlined in the 1987 White Paper. Aerospatiale had teamed up with Canadair of Montreal to fulfill the Canadian Content Value (CCV) of the submission.[39] It was believed by some writers that the link between Aerospatiale and Canadair in Quebec would create a political decision in favour of the *Super Puma* to garner votes in Quebec.[40] But since the policy of Canadian industrial participation was clear from the beginning, EHI had also ensured that it would have Canadian content aboard the aircraft and made partners with Bell Helicopter Textron and Paramax Electronics in Montreal, Canadian Marconi in Kanata, and Industrial Marine Products (IMP) in Halifax.[41] Paramax was actually a subsidiary of the American firm, Unysis Corporation, and had been created in Montreal in order to claim IRB credit for the CPF program. The company was to develop and integrate the mission system for the new ships.[42]

The Canadian avionics industry was designing some of the most advanced ASW systems available. Even before the competition for the NSA contract began, it was clear that Canadian technology would be included in the mission suite of whatever helicopter was chosen. Canadian companies had already been working for two years on individual projects for the possible avionics and systems required for the helicopter. Indeed, one of the objectives of the previous program, the SKR, was to ensure that Canadian industry had the opportunity to participate in the project. As part of the Development Phase in the early 1980s, therefore, the DND and the Canadian government opted to proceed with research and development investments in Canadian industry. According to Stephen Priestley, "this approach may seem back-to-front but ... planners knew that developing first-rate ASW kit would take time. And at the time, Canada had a justifiable reputation as a world leader in ASW technology. Both Canada's Navy and its defence industry wanted to maintain that reputation."[43]

As a result of this investment, four projects were initiated. CAE Industries was awarded a contract for development of an advanced integrated MAD system. Honeywell Canada and Canadian Marconi were awarded a contract to develop the Helicopter Integrated Navigational System (HINS). Computing Devices Canada (CDC) won a contract for approximately $10 million to develop a Helicopter Acoustic Processing System (HAPS). Finally, there was a consortium of CDC, Litton Systems Canada, and Canadian Marconi that was awarded a contract of $32 million to develop a Helicopter Integrated Processing and Display System (HINPADS). In total, nearly $50 million was invested in Canadian industry to facilitate Canadian participation in the implementation of the *Sea King* replacement.[44] Although the airframe to be used was still unknown, the NSA project followed the SKR policy of mandatory Canadian content aboard the aircraft from the beginning. EHI intended, therefore, to maximize Canadian content in the area of ASW avionics and stimulate the developing aerospace industry. In addition, this meant that the *EH-101* would be modified for specific Canadian needs.[45]

Sikorsky was originally expected to make a bid with its SH-60 *Sea Hawk,* in the process of forming the backbone of the USN shipborne rotary wing fleet. The company revealed in late February 1987 that it would not be submitting a proposal because it had determined that its *SH-60* variant would not meet Canada's specifications. Sikorsky had also begun acquiring shares of Westland at the time.[46] The primary companies, and their subcontractors, subsequently attempted to fulfill the rigorous requirements of the RFP, which included expensive material, such as a Global Positioning System (GPS). Regardless of who the winner would be, the first helicopters were expected to become operational in 1995.[47]

Prior to receipt of the two bids, a comprehensive eight-volume evaluation plan was prepared and approved within the DND to describe how the bids would be assessed. This plan was essentially meant to explain how the NSA project office would decide on the bid that fit the requirement and the company that could best undertake the project. That the RFP itself was eleven volumes created the need to plan extensively how to begin to review the bids. The review process would obviously take even more scrutiny. The process was also reviewed by two external consulting companies, Sypher Mueller International Incorporated of Ottawa and E.G. and G. Washington Analytical Services Centre of Rockville, Maryland. An interdepartmental Selection Review Authority and an Evaluation Review Board also took six months to review the bids. The Project Requirements of Military Effectiveness, System Engineering, Program Management, and Support Effectiveness were further broken down into 307 factors, which were evaluated by 88 working groups. There were also the areas of Costs and Contract Terms and Conditions, along with a Collaborative Industrial Program, to be considered.[48]

The level of detail for the NSA procurement was too high. There was effective analysis of possible risks involved in the procurement, which is vital to any acquisition to judge contract compliance. But there was not a clear understanding at the DND that the most dangerous threat to the project was extensive delays. Although there was acknowledgment that some of the software to be used in the aircraft was under development and that there was some "risk in going where others have not tread," the general tone of the evaluation committees was optimistic. In fact, they had concluded that there would be "low risk with EHI developing and implementing a Systems Engineering Program that meets NSA requirements." But part of this analysis should have focused on how long it would take EHI to do it. The process to this point was simply to decide on the model that would enter the next segment, which was really only another Definition Phase called the definition contract. A decision on the final prototype to be purchased that included Canadian technology in the mission system was, in 1987, still another five years away.

The lack of foresight on the part of the NSA PMO was due to a general lack of direction and skilled project management within DND. Each individual involved in defence procurement is integral to the acquisition and maintenance of military equipment for the Canadian Forces. But as A. Crosby, a member of the staff of the Project Management Support Office within the DND in the late 1980s, has written, "project management is, however, poorly understood within the Canadian Forces, within government and in Canadian defence industry."[49] He cited its complexity and lack of uniformity as the primary weaknesses at the time; the procedural manual on project management extended into several volumes, and despite these guidelines each project emerged with an ad hoc form. The absence of an agreed on framework only created confusion. Moreover, the terms used within the DND and between the departments involved were often different. For example, the DND used the terms "Plan," "Develop," "Define," and "Implement" as the four general stages of the procurement process. Because they were not agreed on by all involved – most importantly the defence industry – the communication between them was ineffective and led to a general lack of productivity and defence-industrial co-operation. This situation was underscored by the fact that the government policy was to use domestic industry to expand the capability of the CF for readiness and sustainment.[50]

The report that contained the results of the evaluation of proposals submitted by Aerospatiale and EHI was presented to the NSA Evaluation Review

Board on 2 May 1987. It revealed that in the areas of Systems Engineering, Logistics Support, and Program Management "both proposals were found to be generally acceptable, with certain changes and refinements required during the definition contract phase of the project."[51] But it also concluded that "the key area of concern is air vehicle compliance to specified NSA requirements." One of the factors was "Flight Performance," and the *Super Puma* was rated poor because of concern about engine failure in the hover. Since any new maritime helicopter would be spending significant time in the hover during ASW operations – specifically sonar dipping – this factor was extremely significant. The *EH-101,* with its three engines, was rated excellent in this regard. The *Super Puma* was also non-compliant with the ground-handling requirement, and the evaluation report asserted that Aerospatiale made "unjustifiable claims" that addition of a new larger main rotor, extension of the tail boom, and increase in weight in the Mark II model would not affect its handling. The *EH-101* met the ground-handling requirement. It also met the icing conditions and cabin capacity requirements; the paragraphs on the *Super Puma*'s compliance to these requirements were severed from the document. The strength of the *Super Puma*'s fuselage was also determined to be unacceptable and the weapons system programs poor.

It was determined that the *EH-101* met "all critical requirements," but any attempt to bring the *Super Puma* into compliance would require "either a major re-design ... or a significant change in the operational role of the NSA which would then require a reassessment of the Canadian Navy's concept of operations. A re-design is estimated to cost in excess of $500M."[52] The evaluation methodology decided on in the pre-evaluation phase was a numerical score out of 1,000. The report determined the final grid scores to be

	Super Puma	*EH-101*
Air vehicle	181	838
Avionics	341	564
Military effectiveness	250	720

Clearly, it was not much of a competition. Although the evaluation report comprised eight volumes, the conclusion that only the *EH-101* was compliant was summed up in one sentence at the end: "No level of expenditure will transform the Super Puma into a fully compliant NSA." And, if the bid was this far off, likely the NSA project office knew this before the myriad of subcommittees even began their exhaustive analyses.

Of particular significance is that the report also emphasized that "the proposal by European Helicopter was evaluated as more favourable to Canadian Industry as it provides for a third assembly line in Canada, plus an offer of 7 percent of the value of all EH-101 future production." There was no future export potential of the *Super Puma*. Furthermore, Aerospatiale's proposal of IRBs was focused almost solely in Quebec. As the report concluded, "this is great for Canadair but does little for the remainder of Canadian industry." EHI was determined to have a much more balanced regional distribution of its investment in Canada and create profitable export opportunities. Military capability, therefore, was not all that was being sought by the NSA or the government; domestic industrial investment and the creation of jobs were central to the winning of the NSA contract. The Department of Regional and Industrial Expansion clearly held influence over this procurement, and its evaluators "unanimously consider[ed] the EHI proposal to be notably superior." Although this is another example of politics interfering in Canadian procurement, the selection of the *EH-101* cannot be said to have been solely a political decision. The helicopter was clearly a more capable contender in every way.

On 5 August 1987, Perrin Beatty, the Minister of National Defence, announced that the Canadian government had made a decision to approve funding for a helicopter designed around the *EH-101*. EHI was subsequently awarded a definition contract in April 1988 to begin defining how it would comply with the Canadian requirement of domestic technology within the mission suite.[53] The *EH-101* was chosen because it was a modern helicopter designed for naval specifications, and the NSA management office believed that the *Super Puma* simply could not meet Canada's naval needs because it had been designed as a land support helicopter for the French army in the 1950s. It was reiterated by the management office that the *Super Puma* could meet only 40 percent of the NSA requirement and was not deemed technically acceptable.[54] EHI then had to make a detailed response to the definition contract; this response included covering over 200 specific issues concerning the aircraft, the onboard electronic systems, and shipboard support. It was expected by late 1989.[55] This was an auspicious development for the Canadian Maritime Air Group as the *Sea Kings* struggled to stay in the air during the last months of 1987. One helicopter had ditched at sea off the coast of Halifax on 4 November after an emergency indicator light came on, which signalled a main transmission problem. The aircraft was later recovered after an extensive rescue operation, but the helicopter had essentially been destroyed by landing on the seabed. The next day another *Sea King* above Halifax Harbour also had an emergency light come on, which revealed a possible problem with the tail rotor gearbox; the crew of the aircraft were forced to make an emergency landing in a playing field behind the Herring Cove Junior High School.[56]

The NSA definition contract was not awarded to EHI until April 1988, and EHI delivered an offer on the Implementation Phase for the NSA project in November 1990. This submission exceeded the DND's project budget and was rejected. EHI claimed that it had mistakenly calculated on the department's original requirement, which had been set at fifty-one aircraft. Despite this obstacle, the *EH-101* was still thought to be "the only aircraft, in production or development, capable of fulfilling the critical aspects of the operation requirement."[57] Moreover, the NSA project was an essential component of the overall naval refurbishment program, "without which the substantial previous investment in the Canadian Towed Array Sonar System (CANTASS) and the Canadian Patrol Frigate (CPF) will not yield an effective surface and sub-surface surveillance capability." A modern helicopter was expected not only to be the eyes and ears of the ship but also the rapid response vehicle in an emergency SAR situation or medical evacuation. It was necessary, therefore, to overcome any problems that would impede implementation of the NSA project. Public opinion at the time also demanded that replacement of the *Sea Kings* be accelerated, as one had crashed off the coast of Bermuda on 19 September 1989. As one editorial read, "now it is time for them to join the vintage machines that line the main entrance to Shearwater and will never fly again."[58] The article also explained that the *Sea Kings* in use by the British and American navies were newer versions than that used in Canada and were much more advanced in ASW technology. Indeed, the British were still building new versions of the *Sea King*. The *Mk 6* was assembled in April 1987 and included Very High/Ultra High Frequency (V/UHF) and an integrated acoustic-processing system, the AQS-902G.[59] Martin Shadwick later explained how the British had been far more aware regarding the life cycles of their helicopters than the Canadians:

> In the British case, the Sea Kings they are updating are quite youthful compared to our Sea Kings which were all delivered between 1963-1968, whereas the British have large numbers of Sea Kings that were delivered in the late 1970s and on into the 1980s. So, the optics are a bit different, and the interesting thing is, they're updating Sea Kings that are a fraction of the age of ours, and still planning to buy the EH-101, which I find fascinating.[60]

While it was obviously much easier to sell maritime interests to the British due to their long naval tradition than it was to Canadians, it was understood at NDHQ that it was time to make progress on the NSA.

The turbulent events of November 1989 with the fall of the Berlin Wall and the subsequent collapse of the Soviet Union meant that *Sea King* operations were about to enter a new era. Although it was originally expected that there would be a relaxation of world tensions, the end of the Cold War seemed to lift many of the bipolar constraints on open conflict. States then focused on local hostilities – either within their borders or with a neighbouring state.

As international hostilities flared up, the helicopter squadrons found themselves increasingly called on to commit resources to subdue or prevent new clashes. As the "History of 423 Squadron" put it, "aircraft and crews were expanding beyond the traditional HS roles, and coming in harm's way more often."[61] The most significant of these operations occurred after the invasion of Kuwait by Iraq in August 1990.

On 10 August, several helicopter air detachments (HELAIRDETS) were busy preparing for Exercise Teamwork – a major North Atlantic Treaty Organization (NATO) exercise in the eastern Atlantic – when Prime Minister Brian Mulroney announced that Canada would be sending *Athabaskan, Terra Nova,* and *Protecteur* to the Persian Gulf to help enforce United Nations (UN) resolutions against Iraq. The *Sea King* detachments that were to travel with them would be used for surface warfare:

> Six helicopters were stripped of their Anti-Submarine Warfare [ASW] gear and fitted with self protection equipment and surface surveillance gear. These fits included Forward Looking Infra Red (FLIR) and Night Vision Goggles (NVG), used to see at night; GPS navigation equipment; an Infra-Red jammer and flare dispensers to protect against heat seeking missiles; a radar warning receiver and chaff dispensers to confuse radar guided missiles; and a light machine gun mounted in the aircraft's cargo door ... The engineers and technicians were condensing years of work into a mere two weeks.[62]

As already mentioned, chaff and flare protection had been added to the USN *Sea Kings* in 1971.

Since the Canadian *Sea Kings* operated in a cold maritime environment, extensive modifications had to be made to their electronic, navigational, and crew survivability for use in the high-temperature, dust-filled environment of the Arabian Desert.[63] The crew were also outfitted with nuclear, biological, and chemical defence (NBCD) suits and personal body armour.[64] On 24 August, five helicopters and their ground and air crews embarked with the ships and headed for the Persian Gulf. The sixth converted helicopter remained in Shearwater to continue testing of the new equipment and development of procedures and tactics. Once it arrived in the theatre of operations, 423 Squadron began operating in newly defined roles. One pilot explained that,

> Because of our equipment fit, we were tasked to fly our operational sorties of interdiction (of shipping) mostly at night. With FLIR and NVG's we were able to identify all shipping at night. We hailed vessels to determine the ship's cargo and destination. Any suspect vessels would be boarded and searched and then diverted or allowed to proceed. While there, we developed

a method of inserting a boarding party using the helo, called VISIT (Vertical Insertion Search and Inspection Team).[65]

In addition to interdiction patrols, the Canadian squadron was involved in mine search and anti-ship missile defence after the war officially broke out. The brinkmanship in the gulf caused a significant increase in the flying rate of the *Sea King*. This led to increased stress on the airframe, and the age of the *Sea Kings* quickly became a concern. As Peter Charlton and Michael Whitby have written,

> Metal cracks in the airframe were a constant worry, primarily because of the effort and logistics required to repair them and the consequent negative effect that they could have on overall aircraft availability. The Sea Kings were getting old and brittle, and the constant strain on the airframe brought on by corrosion and inherent vibration made them prone to this kind of damage.[66]

The ages of the helicopters had also made them increasingly difficult to maintain. Eighteen missions either were aborted or returned prematurely due to a maintenance problem. Major mechanical repairs such as rotor head and gear box failures that were normally conducted on land to ensure steadiness of the aircraft had to be conducted onboard the rocking ship. Despite these setbacks, there was tremendous confidence in the air technicians to maintain airframe hours, and Canada performed well in Operation Friction despite the use of a helicopter that was not designed for desert conditions.[67]

In the aftermath of the war, it was stated by Maritime Command that future operations in the region might involve an embargo and a blockade. The Arabian Sea and the Persian Gulf area would involve possible engagement of sophisticated submarines and fast patrol boats equipped with modern long-range anti-ship missiles. The sensors of a modern helicopter would be essential in this role, including its ability to dispatch boarding parties onto any vessel attempting to run the blockade.[68] Colonel John Orr has stated that "a renaissance of sorts occurred with the first Gulf War when the Sea King took on an increasing role in surface warfare (principally surveillance) and utility operations. This change in emphasis away from ASW continued throughout the new century."[69] But clearly there was also a realization that submarines would be a part of future operations.

Since EHI had been selected as a sole source for completion of the NSA, it was expected to deliver all aspects of the project, which included the airframe and the mission suite. The approach of having a single prime contractor with "Total System Responsibility" had been the norm for DND major crown projects since the 1970s.[70] In such cases, however, the DND was bound to ensure that it had a sufficient budget to cover any incremental costs associated with this approach. These included any costs related to the value that the prime contractor assigned to the risk it assumed for the project and any "marking-up" it added to the final project deliverables produced by its subcontractors. The DND and the NSA project office, therefore, did not think that they could responsibly undertake the first implementation offer made by EHI as they anticipated major cost increases. As a result, the three shareholders of EHI developed an "Alternative Program Plan" that was delivered to the crown on 15 March 1991. The three shareholders included Agusta, Westland, and, as of September 1989, Unisys Corporation of Pennsylvania, the parent company of Paramax. Unisys came in to restart the electronics development after there were delays and disagreements between the other two companies on the matter.[71] The salient features of the new proposal included:

(1) making the NSA more common with the air vehicle being provided for similar British and Italian navy programs in order to reduce nonrecurring engineering costs;
(2) incorporating a reduced equipment baseline to lower costs without unacceptable degradation of operational capability;
(3) undertaking industrial reorganization to reduce corporate layering and its associated costs; and
(4) combining the NSA project and the procurement of new SAR helicopters, changing the total number of *EH-101s* to be acquired from thirty-five to fifty.

Section 2 of the proposal meant that part of the electronic warfare package was removed, which meant that the aircraft could detect whether the submarine being tracked had locked on to it and was aware of its approach; this would have made it more effective at bearing down on the target.[72]

The implications of section 3 were that the shareholders suggested a two prime contractor approach to implementation. In this scenario, Agusta and Westland, through EHI, would contract to deliver the basic helicopters, while Unisys, through Paramax, would contract to integrate the mission equipment, to deliver the complete "prime mission vehicles," and to define the "Integrated Logistics Support Package." In this approach, the Canadian government would supply the basic aircraft as "Government Supplied Materiel" to Paramax, and it would subsequently assume complete responsibility for the satisfactory performance of the whole aircraft at delivery. As of November 1991, the project office had examined the proposal and believed that it had "the potential to be a practical and financially viable approach to the implementation of this project."[73]

Section 4 of the proposal was to play a pivotal role in the quest for completion of the NSA project. The DND's fleet of SAR helicopters, the CH-113 *Labrador,* was by that time more than thirty years old and was expected to reach its extended life expectancy in 1998. The New Search and Rescue Helicopter (NSH) project office was created by the DND in 1988 to acquire a replacement helicopter, and in December 1990 an options analysis paper was presented to the DND Program Control Board for consideration. This was essentially done at the same time that the first proposal by EHI had been submitted and rejected. The options analysis concluded that

(1) the *EH-101* was the best overall solution, in terms of operational capability and supportability, to the NSH requirement;
(2) during the in-service life cycles of the NSA and NSH, the "cost avoidance" achievable through the use of a single aircraft type was a minimum of $275 million; and
(3) there were additional savings achievable during the capital acquisition stage of the two projects through the economies of scale, which were at least partially dependent on the co-implementation of the projects.[74]

The Program Control Board of the DND accepted the conclusion that the *EH-101* was also the best helicopter for SAR, and therefore, after considerable study, it directed that the NSA and NSH projects be pursued as a joint venture. This move created controversy as the other major helicopter manufacturers, such as Boeing, Sikorsky, and Aerospatiale, were not invited to make bids. It should be noted, however, that Boeing had already rejected accepting any RFPs for new contracts in April 1990, as "current facilities were taxed to their limit," and Sikorsky had not shown any interest in competing for the SAR contract.[75] Furthermore, Aerospatiale's *Super Puma* had already been determined to meet only 80 percent of the NSH requirement.[76]

Boeing actually made an informal proposal to the government in July 1992 as an alternative to the NSH portion of the project. Although it admitted that it could not fulfill the naval requirement, Boeing believed, by that time, that it could provide a modernization to the *Labradors* and that this would be a more economical pursuit for Canadian SAR. This option was subsequently discussed in the hearings of the Standing Committee on National Defence and Veterans Affairs (SCONDVA) in May 1993. Although those on the committee opposed to the NSA/NSH project wanted to use the significantly lower price that Boeing had offered for a modernization as evidence of Conservative mismanagement, it became clear that the actual capability offered on an upgrade of forty-year-old helicopters was not acceptable. This was best elucidated by the following exchange on the capabilities of an upgraded *Labrador* and an *EH-101*:

> Conservative MP Robert Hicks: Surely, we have to begin by admitting that the EH-101 has tremendously greater capabilities and multi-role capabilities than either an upgraded Chinook or a Labrador. Would you agree with that, sir?
>
> George Capern (Vice President, Government and Industry Affairs for Boeing): I cannot debate that and we don't want to debate that. We have made a practice of doing that.
>
> Hicks: Okay. Are you willing to offer an opinion as to how wise it would be for Canada to continue with helicopters that might become 60 years old after they have been worked over a few times or Canada continuing to use three different types of helicopters which means three types of servicing, three types of technicians, three types of pilot training, three types of service training, and three reservoirs of spare parts?
>
> Capern: The only observation I would make, Mr. Hicks, is that we do not regard any of our equipment to be multi-role.[77]

Clearly, the Boeing scheme did not comply with the vision of the DND and the government of a "rationalized fleet," which meant one machine for many roles. Although the government would save money by merely upgrading and deferring replacement of the *Labrador,* the capabilities lost would have been vastly disproportionate.

Canada's topography, geography, and variable weather conditions made it one of the most challenging environments in the world for search and rescue operations; the new SAR helicopter was expected to rescue survivors on the seas, from the decks of ships and oil rigs, from mountainsides, isolated arctic islands, and inland disaster sites. The range, speed, endurance, and anti-icing capability of the *EH-101* made it the perfect candidate.[78] A recent example highlighted the need for a modern SAR helicopter; when a *Hercules* aircraft crashed in the Arctic near Alert on 30 October 1991, the helicopter transported to the area was largely restricted by weather and was not able to carry out a timely SAR operation. Some of the survivors of the crash later died from injuries and the cold. It was speculated that the *EH-101* could have reached the crash site in time to avoid this unfortunate result.[79]

EHI was subsequently advised of the sole source direction that the DND wanted to take to combine the projects, and the company's Alternative Program Plan was set out to include fifty aircraft – thirty-five of the NSA variant and fifteen of the NSH. Although this plan was approved, it still remained for EHI and Paramax to deliver a final offer to the government for contract consideration. The initial plan, however, was very optimistic; it identified an outright saving of $150 million associated with the combined project. This saving was attributable to factors such as a single contractor management team, a single uninterrupted production run, and a combined project office. Clearly, a reduced number of aircraft models within the Canadian military could result in further life cycle benefits as there would be no need for two training systems, two simulators, two maintenance systems, two sets of spare parts, or other redundant facilities and equipment. In November 1991, it was anticipated that there would be a submission for effective project approval in the spring of 1992 and that it would include the second variant of the *EH-101* for the SAR role. The joint project was considered to be "affordable based on a revised acquisition strategy involving dual prime contractors."[80]

By the late 1970s, Naval Staff discussions began on the replacement of the *Sea King*. Not only was its airframe near the end of its service life, but also, within the next decade, its internal mission systems would be outdated as the Cold War continued to progress and submarine warfare technology evolved. As the defence policy continued to stress Anti-Submarine Warfare, Canadian naval officials remained committed to its effectiveness in the field. The joint ship and helicopter platform remained integral to national sovereignty and collective defence, and replacement of the *Sea King* was necessary to carry out this government mandate. Much like procurement of the *Sea King* itself, however, a replacement would elude naval procurement officials, and the interim solution of a modernization program was carried out instead. Although discussions to replace the fleet did not come to an end, the bureaucratic process needed to identify the capability deficiency, justify a new requirement, and analyze the available options made little progress until the mid-1980s. The *Sea King* Replacement Program was renamed the New Shipborne Aircraft Project in 1981, and though it was registered in the Capital Defence Program that year it was not approved by the Defence Management Committee (DMC) until 1985. The NSA sought to introduce a multi-role capability into the fleet to work in tandem with the CPFs under construction and the 280 *Tribal* class destroyers that were to be upgraded. The final Statement of Requirement, however, was not ready until 1987.

In 1987, after an exhaustive process, the *EH-101* ASW helicopter was formally selected to fulfill the Canadian NSA role. It was widely considered the most capable of all the existing naval helicopters, even compared with those under development. But its selection signalled only the beginning of another Project Definition Phase; the NSA definition contract was not awarded to EHI until April 1988. One of the objectives of the contract was to define in detail the Canadian business and industrial benefits that were officially required by the Canadian government for the subsequent NSA implementation contract. It still remained for EHI to establish, therefore, how it would comply with the 1987 White Paper and stimulate Canadian industry. Since this compliance would eventually include hundreds of Canadian companies, it caused the NSA to enter an indeterminate state caused by seemingly endless negotiations. By 1989, a *Sea King* crash highlighted the calls to replace the aircraft as soon as possible. The first proposal by EHI, however, did not come until three years after the *EH-101* had been selected. Its first proposal was immediately rejected because it was believed that the risk of escalating costs was too high. Matters became increasingly complicated as near-simultaneous discussions were carried out to decide whether a version of the *EH-101* could also fulfill the Canadian NSH requirement. The NSH was subsequently merged with the NSA, and the Contract Definition Phase had to be restarted.

EHI was awarded a definition contract for a combined NSA/NSH acquisition on 30 December 1990, and the new requirement was set for thirty-five NSA and fifteen NSH. Although this plan was expected to create a significantly lower price overall, it initially failed in this capacity, and EHI had to pare down many of its aircraft components. The new proposal was given to the government in March 1992, and it was accepted in September. Since the first contract had been signed and was almost complete, there was no thought that the program was vulnerable to political attack. The NSA program was part of the public record, and there had never been any tangible opposition to fulfillment of the naval helicopter role in Canada. Negotiations continued, therefore, on the terms of acquiring the *EH-101* for the naval and search and rescue roles. A major problem, however, was that, by putting the two projects together, this created one highly expensive program instead of two moderately expensive programs. This may have been an issue of semantics to some, but expenditures on defence are very sensitive in Canada. Inclusion of the NSH in the NSA was a wise decision on many levels and likely would have brought the NSH program to fruition more quickly than expected. It would have the opposite effect on the NSA program, however, which had already been officially under way for seven years.

EH-101 Maritime Helicopter.
Reproduced with the permission of AgustaWestland

6
The Vulnerability of the NSA: Political Parrying

> Across Canada, the *EH101* project will create 45,000 direct and indirect person-years of employment over 10 years. Canadian companies will supply a minimum of 10 percent of the *EH101* airframe and 83 percent of the electronics systems for the CF helicopters. All told, the Canadian *EH101* will be more than 50 percent Canadian-made ... and not to be overlooked is the guarantee that 10 percent of every *EH101* sold worldwide will be Canadian-made ... Perhaps the best part of the deal are the technology transfers to Canadian suppliers.
>
> – Kim Campbell, Minister of National Defence, February 1993

> This issue has been studied, studied and studied. Any rational and logical thought process leads to one conclusion, replacement of the current fleet as it nears the end of its life and the replacement of the current fleet with the EH-101, all weather, all purpose aircraft is a must.
>
> – Paul Dick, Minister of Supply and Services, presentation to the Standing Committee on National Defence and Veterans Affairs, 13 May 1993

By 1991, there were many players involved in the seemingly imminent replacement of the *Sea King*. One *aide-mémoire* prepared by the Ministry of Supply and Services to provide other ministers with information on the current status of the New Shipborne Aircraft (NSA) was circulated to the Ministers of Industry, Science, and Technology Canada, the Atlantic Canada Opportunities Agency, Western Economic Diversification, the Treasury Board, External Affairs and International Trade, and the Department of

Finance.[1] The growing number of agencies and ministries involved created competing priorities and slowed the procurement process considerably as each concern had to be taken into account before moving forward. The NSA project office had been directed by the government to promote long-term regional industrial development and assist Canadian firms in becoming competitive in domestic and world markets. The operational requirements of the military, therefore, were mixed within a system of competing priorities. Although it is true that Canada already had a solid reputation in ASW technology, much of the electronics to be used in the *EH-101* had yet to be designed or contracted for. Computing Devices Canada (CDC) was designing the Helicopter Acoustic Processing Systems (HAPS). A joint development project between CDC, Canadian Marconi Company, and Litton Systems Canada was to produce the Helicopter Integrated Processing and Display System (HINPADS). The Tactical Co-Ordinator (TACCO) and the Sensor Systems Operator (SENSO) would use the HINPADS to format and then transfer tactical command data to the large display screen in the centre of each of the two pilots' instrument panels. And CAE Electronics had already designed the Advanced Integrated Magnetic Anomaly Detection System (AIMS), considered a likely addition to the *EH-101* at the time. The aircraft would also need a 360-degree radar, a tactical data link through satellite, a Missile Approach Warning System (MAWS), countermeasures equipment, electronic support measures (ESM) to monitor frequencies against possible surface-to-air missiles (SAMs), and infrared equipment for night operations, to name just a few more components, some of which the NSA did not yet have contractors for.[2]

The complexity of creating this new mission system in Canada and then installing it into the *EH-101* airframe was obvious. These realities increased the risk factor for the project. This situation occurred even though European Helicopter Industries (EHI) had offered the mission suite and avionics already purchased by the Royal Navy (RN). The responsibility for ensuring that each item would be synchronized into an overall system was that of Paramax and Canadian Marconi. It was a staggering technological task, and a multitude of subcontracts were expected to be involved. These negotiations and subsequent contracts, along with the manufacture and installation of them into the aircraft, were to add a significant amount of time to the final procurement schedule. The final helicopter would then have to be certified.

Another factor that had already delayed the overall program was that the Department of National Defence (DND) did not want to be the first to procure the *EH-101*. The British were interested in the *EH-101*, and prototypes had already been created for them, but they had not signed a contract yet. If Canada were to sign before them, this would have meant accepting the first production line, and the DND had already experienced problems with such an undertaking with the procurement of the New Fighter Aircraft (NFA) Program of 1980. The NFA project continued the policy of linking procurement to Industrial and Regional Benefits (IRBs), and because Canada was to be the first major foreign customer the government was able to negotiate a more extensive benefits package. The Request for Proposal (RFP) included an entire volume on "Canada's expectations and requirements in this area."[3] The project also followed the format of having a Contract Definition Phase and had the best two potential contractors compete to see who could provide the best offer. The IRB proposals of the two bidders played a vital role in the selection of the *F-18A* by McDonnell Douglas over the *F-16* by General Dynamics.[4] The first models received were originally calculated to have 6,000 flight hours available; however, once they were received, it was discovered that they would be able to fly for only 2,000 hours. The first models produced, therefore, were two-thirds less capable than the later models. As former Assistant Deputy Minister (Materiel) (ADM Mat) Ray Sturgeon explained, "given the previous experiences, DND senior officials decided that they would not enter into a contract for the first production run of a new weapon system."[5] This problem was solved in October 1991 when Britain's Ministry of Defence ordered forty-four ASW versions of the *EH-101,* known as the *Merlin HM Mk I.* A further twenty-two aircraft of the utility variants, known as the *Merlin HC Mk3,* were ordered for the Royal Air Force (RAF).[6] The Sikorsky SH-60F *Sea Hawk* had already been chosen to replace the United States Navy's (USN) SH-3H *Sea Kings* for ASW in 1987.[7] The first came into service in 1989, and the transition was completed in the mid-1990s.

The industrial benefits to Canada of the *EH-101* acquisition were advertised as being extensive. They were expected to total 113 percent of the purchase price; this meant that more money was expected to flow into Canada over the lifetime of the project than the government was to spend on it. These benefits would be spread out across Canada. The contractors indicated the value of the work to be approximately $3.236 billion. The regional distribution of the benefits was judged to be approximately $523 million to the Atlantic provinces, $975 million to Quebec, $874 million to Ontario, and $726 million to western Canada, with $138 million yet to be allocated.[8]

In March 1992, EHI and Paramax delivered their revised offer to the government, and it was approved by the Defence Management Committee (DMC) in June. That July the defence minister, Marcel Masse, astutely pointed out that the government had to make progress on the program or it would become "more and more difficult."[9] Indeed, there was a feeling that, if cabinet approval did not come soon, the money set aside for the project would vanish through continued budget cuts. It had also became obvious to many that defence procurement had to be simplified, and in a document entitled *Canadian Defence Policy 1992,* the government gave clear indication that it would do so in part by focusing on off-the-shelf equipment and "avoiding unique Canadian solutions that require expensive and risky research, development or modification of existing equipment."[10]

Canada had also entered a recession in 1989, and the Mulroney government, as Dan Middlemiss has written, "eviscerated its own 1987 White Paper plans to redirect, rebuild, and revitalize the Canadian Forces (CF)."[11] Indeed, the April 1989 budget cancelled, reduced, or deferred expenditures for many of the major equipment projects that were planned. The gross domestic product had dropped, and unemployment rose from 7.2 percent in 1989 to 11.2 percent in 1992.[12] On 12 November of that year, Mulroney admitted that he had misjudged the severity of the recession.[13] Although the government had acknowledged the "commitment-capability gap" in 1987 and the impending "rust-out" of CF equipment, it was no longer in a position to provide general relief. Despite the serious fiscal restraint that prevailed, the NSA was spared. The Conservative government knew that the helicopters had to be replaced and was determined to follow through on its commitment.

Masse also had to counter arguments that the project would be a political gift to Quebec.[14] He insisted that he had not finalized where all the regional benefits would go and would only say that Montreal just happened to have a large industrial base in electronics and aerospace. The initial accusations in the newspapers of patronage were justified as the government had just given approval for a $1.293 billion contract, without competition, to Bell Textron in Montreal for utility helicopters for the army.[15] In fact, there was a long history of defence contracts going to Montreal to garner votes in Quebec; Wilfrid Laurier had ensured that the manufacture of the Ross rifle before the Great War went directly into his riding.[16] The most recent and perhaps most infamous was the maintenance contract for the *CF-18*. After procurement of the *Aurora,* Premier of Manitoba Ed Schreyer criticized the federal government for giving a major subcontract to Canadair in Montreal and argued for a larger share of aerospace work to be done in his province. In 1978, the Manitoba government created the New Fighter Aircraft (NFA) Task Force to lobby the federal government and the contractors involved for a 10 percent share of the project's IRBs. The Manitoba lobby could not compete with that of Quebec, however, as the latter had lost the main portion of the IRBs for the main contract to Ontario-based McDonnell Douglas. Quebec was already in the middle of the sovereignty-association referendum, and relations with the federal government were strained. As a result, the federal government attempted to assuage Quebec by giving it the maintenance contract. It was awarded to Canadair on 31 October 1986. This decision went against the advice of the NFA project evaluation team that a better bid had come from Winnipeg's Bristol Group.[17] Despite these precedents, the NSA project could not be labelled a patronage procurement for Quebec as it had already been through extensive analysis and competition going back to 1987.

The total project cost for the NSA/NSH (New Search and Rescue Helicopter) was $4.4 billion. Cabinet gave formal approval for the program in September, and on 8 October 1992 a contract was signed jointly with EHI and Paramax to supply fifty fully integrated *EH-101* helicopters at a combined contract cost of $2.8 billion. A contract of $1.4 billion was signed with EHI for the manufacture of fifty airframes, with General Electric Canada to supply the engines. Another contract for $1.4 billion was signed at the same time with Paramax for the development and manufacture of the electronic equipment to be installed in the thirty-five *EH-101*s to be used in the maritime role. Finally, there was an additional $1.6 billion for training, spares and contingencies, project management, and ship conversion. Approximately $90 million was already owed to EHI for its detailed solution to the original stated requirement during the Design Definition Phase.[18] It was to be one of the largest defence procurement programs in Canadian history, and approximately 400 companies were expected to receive work from it.[19] The following week Conservative ministers travelled to various regions of Canada to explain how they would benefit from the contract. It was not just Quebec that was expected to gain from the venture.

Until June 1992, there had been little public debate on the NSA except in military, industrial, and academic circles.[20] It was public knowledge that the DND had been preparing to replace the *Sea King* since 1986. But then news stories began to circulate explaining that cabinet was close to final approval of the purchase of the *EH-101*. The contract suddenly became a focal point of attack for the Liberals and the New Democratic Party (NDP), which questioned the need to replace the helicopters and argued that modernizing the fleet again would be a cheaper alternative. The new Bloc Québécois was also opposed to the program. The Liberal plan was to upgrade the *Labradors* and the *Sea Kings*, a project determined by the Minister of Supply and Services to cost $600 million and $1.8 billion respectively. Another *Sea King* modernization only deferred the need for replacement of the aircraft and was ultimately more expensive. The plan seemed even less palatable when joined with the fact that the NSA project office had already determined that no amount of upgrading would meet the navy's requirement for a modern helicopter.

NDP Members of Parliament (MPs) indicated that, if their party formed the next government, the helicopter purchase would be cancelled. The Liberals, while strongly criticizing the purchase, claimed that they would review it if they were elected. The only other party to support the NSA/NSH was the Reform Party. It was concerned, however, that there was not enough reliable information available to the public and that decisions would be based on "hearsay, rumour, and ignorance."[21] And so, by June 1992, as one newspaper article put it, "what had been a straight-forward and well publicized six-year procurement plan to replace the helicopters has turned into an 11th hour political hot potato."[22]

Defence officials responded to these early threats in publicly available sources by claiming that it was simply inconceivable to cut a program that was both necessary and already years under way. One analyst observed that,

> As to the unthinkable alternative of cancelling the NSA program, apart from the loss of money already spent or committed plus cancellation charges, it must be recognized that the new frigates are reckoned to be only about 60 percent effective without helicopters. Apart from this, the NSA is still regarded by DND as a sacrosanct program that is essential to carry through.[23]

Internally, DND officials also began to defend the NSA. A document by the DND in 1992 noted the continued importance of the project:

> One of the main reasons our vintage ships have remained productive to the end of their frequently extended lives is, in fact, because of the effectiveness of our ship and helicopter team at sea. Our vintage anti-submarine warfare ships are being rapidly replaced by Canadian patrol frigates and modernized Iroquois [*Tribal*] class destroyers which fully satisfy today's general purpose fleet requirements, but one key element of the naval team, namely the Sea King helicopter, can no longer do so, and our new ship/old helicopter combination is becoming less effective.[24]

Part of the task of the navy within the national defence policy, as stated in 1987, was to provide national sovereignty through surveillance of Canadian shores and waters. The thirty-five NSA were to be located on both the East and the West Coasts to provide operational support to the navy's two task groups. Each group was to consist of up to five CPFs, one *Tribal* class destroyer, and one Auxiliary Oil Replenishment (AOR) ship. Typically, a single aircraft or single ship with its helicopter would be responsible for establishing a Canadian presence in a determined area. A single ship conducting a sovereignty patrol of this kind could continuously cover an area of approximately thirty miles, the maximum distance of its radar. A shipborne helicopter, fitted with a modern radar and flying at a height of about 6,000 feet, could monitor an area of 100 miles. Although this added an obvious increase in surveillance potential, the *Sea King* was still using twenty-year-old tube technology radar, which covered four times less area than did the equipment of more modern helicopters. The fact that the *EH-101* also had a maximum speed of almost 300 kilometres per hour, had a range of

550 nautical miles, and could remain in the air for over four hours made it the perfect machine for these tasks.[25]

A brief prepared by Maritime Command and Maritime Air Group Headquarters expanded on the *EH-101*:

> The spectrum and sensitivity of the aircraft's sensors will play a significant role in supporting the interdiction of drug smuggling and illegal immigration. Day and night, in almost any weather condition, the aircraft will be able to monitor our ocean boundaries and will be instrumental in the protection of our resources and sovereignty. Fishery patrols and environmental surveillance will reap an immediate benefit in the fact that the reporting and sampling data acquired by the aircraft will be of sufficient accuracy to withstand judicial scrutiny.[26]

It explained that helicopters were also vital in high traffic areas, such as heavy shipping, where the aircraft could track several targets at once and relay the information back to the parent ship. But where the *EH-101* could track thirty targets and then send the information back to the ship instantly through a data link for surface picture analysis, the *Sea King* did not have this ability. Since the *Sea King* had to rely on passing information on by voice, no more than five contacts could be tracked and assessed at once. Only eight of the *Sea King* helicopters had Forward Looking Infrared (FLIR) in 1992, so any identification at night was completely dependent on having one of those helicopters available. Another important feature that the *Sea King* lacked was modern anti-icing equipment, which was necessary for the helicopter to operate for one-third of the Atlantic winter period.[27]

The second official fleet priority of the navy since 1987 was contributing to collective defence, and this priority had not changed in 1992. Naval officials, therefore, continued to stress the necessity of carrying through with the NSA project. A DND document noted that, "whether the commitment involves a NATO or a bilateral Canada-US operation, the numbers, duration, complexity, and distance from home port always result in a significant increase in support requirements when compared to those for national sovereignty operations."[28] For example, when a helicopter-carrying destroyer (DDH) was assigned to the Standing Naval Force Atlantic,both vehicles had to be prepared for a six-month deployment overseas. To keep them operational, the document explained, "one must have robust equipment, extensive on-board maintenance facilities, and a large spare parts inventory."

The DND document also revealed that the integrated logistics support aspect for any replacement helicopter could not be reduced unless Canada was willing to sacrifice the capability to undertake long-range, extended deployments. Such deployments also involved more than just finding a contact at sea; a modern helicopter had to use its sensor and data link technology to coordinate effectively any attack with the CPF's Harpoon missile. Although a ship was capable of detecting a submarine at ranges in excess of 100 miles, it could

> deliver a ship-launched anti-submarine homing torpedo to a range of only five miles. The Canadian Navy therefore relies on the helicopter to resolve the submarine's exact position, to identify the submarine as friend or foe, and ultimately deliver the attack. The fact that the modern submarine is itself armed with powerful ship killing torpedoes and missiles but cannot attack an aircraft also lends a tremendous attraction to having the helicopter, and not the ship, close for the attack.[29]

The new replacement helicopter clearly had to be capable of modern electronic warfare and required a good radar detection and identification capability. The *Sea King* had been designed for a less complex theatre of maritime warfare and lacked these essential capabilities.

The DND had also requested that the new helicopter have a multi-role capability.[30] This requirement was made obvious in the 1990 Gulf War; the conflict in Somalia was another example of the changing international security environment. Her Majesty's Canadian Ship (HMCS) *Preserver* was called to Somalia on Operation Deliverance in the fall of 1992. By 13 December, the ship had deployed with four full crews and arrived in Mogadishu. The helicopters were pressed into service immediately, with the primary task of transporting supplies ashore in support of the United Nations (UN) efforts to relieve a country ravaged by civil war. The detachment would ultimately transport 430 tons of supplies and fly over 520 hours on the UN-authorized mission. The maritime helicopter, therefore, proved to be an ideal vehicle for moving personnel and cargo among ships in a task force or to a shore facility. It was also used in reconnaissance for the Canadian Airborne Regiment, US Marines, and US Army Rangers. These flights included overland night surveillance using the FLIR capability for gathering intelligence. The missions to seek out arms caches and rogue Somalis were not without danger. The *Sea Kings* occasionally encountered heavy fire. The helicopters were also involved in medical evacuation; crews completed so many of these flights that the senior USN medical officer affectionately called them the "Body Snatchers."[31] The Canadian helicopter capability demonstrated a great deal of flexibility in the new security environment, and this adaptation to new roles was precisely why the military wanted a modern maritime helicopter. It was obvious that helicopters would continue to play a role in supporting Canadian operations around the world after the Cold War.

As for interoperability, DND officials believed that, if Canada was going to continue the policy of collective defence, then its equipment had to be able to synchronize efficiently with that of its allies. The most obvious aspects involved fitting the helicopters with interoperable radios and fuelling systems to allow the allies to work together. The *Sea King* lacked these options.

Finally, a modern helicopter would allow the navy to continue to carry out the secondary roles of the aircraft. By 1993, they involved resupply; communications relay; Search and Rescue (SAR); medical evacuation within an overseas task group; and vertical replenishment, which involved transporting materiel and people to wherever they needed to go in the theatre.[32] Regarding medical evacuation, the *EH-101* could carry sixteen stretchers or thirty people; the *Sea King* could carry four stretchers or ten to twelve people.[33]

In July 1992, a DND backgrounder stated that the NSA is "a vital purchase that is fully consistent with Canada's defence policy. This policy enunciates the Government's commitment to maintain flexible capable armed forces for the defence of Canada and Canadian interests both at home and abroad in the face of an array of threats, challenges, and problems that Canada can be expected to face during the next two decades."[34] The cost of maintaining the *Sea King* fleet until it was replaced was escalating and was estimated to be $46,307,200 a year until 1995-96.[35] Further estimates were not carried out by naval officials as they expected to have the new aircraft by that time.

All available options for dealing with the *Sea King*'s deficiencies were analyzed and put forward by the NSA project office.[36] Although the consortium of AgustaWestland had won over its primary competitor in 1987, Aerospatiale, the NSA project office continued to compare the *EH-101* to some "conceptual aircraft" that had been designed by 1992: the *S-92* by Sikorsky and the *NH-90* by NH Industries. The NSA requirements were based on "Sovereignty, Surveillance, Fisheries Patrols, [and] Naval Operations."[37] It was determined that the *EH-101* fulfilled 95 percent of the NSA requirements, whereas the *S-92* was believed to meet only 75 percent and the *NH-90* only 10 percent. The newer version of the *Sea King* was also placed in the analysis and met only 20 percent. It was concluded that the other two existing models, Sikorsky's SH-60 *Sea Hawk* and Aerospatiale's *Super Puma,* satisfied 55 percent and 40 percent respectively. One of the *Sea Hawk's* main challenges was that it had a "limited size," and the new *Sea King* model had an "incompatible mission suite." It was decided in 1992, after five years of study, that the *EH-101* remained "the best choice for DND." Although the differences in price for all the possible choices were relatively small, the level of capability of the *EH-101* over its competitors was enormous. It was also concluded that EHI was a company of high standards and could develop economic potential in all regions of Canada.[38]

The defence of the NSA put forward by the National Defence Headquarters (NDHQ) and the Conservative government was categorically rejected by the Liberal Party. On 2 July 1992, Jean Chrétien expressed his disapproval of the NSA purchase because the threat of the Cold War had ceased to exist. He stated in a letter to Paul Manson, President of Paramax, "in closing, I wish to assure you that the Liberal Party is fully committed to upgrading the equipment of the Canadian Armed Forces ... But our procurement priorities must reflect the real security needs ... of the Canada of today – not the Canada of twenty years ago."[39] Manson had been the project manager in the acquisition of the *CF-18* and was an authority on Canadian procurement by that time. In his response, he stressed the area that he knew was most important to politicians – that of regional economic benefits. He explained that Canadian companies had been selected by EHI to build the electronics mission system not only for the Canadian order but also for all orders worldwide. The international market forecast at the time was 800 aircraft. Manson declared that, "as a result of these arrangements, some 50,000 person-years of highly skilled work will be produced by the program, distributed equitably from one end of Canada to the other."[40] He also pointed out that, though it was true that Canada was currently in a recession, most of the money for the NSA/NSH would not be needed until near the second half of the 1990s; expenditures in the first three years would be under $400 million a year. Manson ended the letter by expressing hope that there would be further study of the project and included four pages of quotations praising the IRBs that would come from fulfillment of the contract. Not even two weeks later Chrétien released a statement claiming that he and his party supported the modernization of the navy's helicopter fleet and replacement of the *Sea Kings* but that unemployment was a bigger issue: "Our biggest problem now is the economy, not defence. Let's deal with the economic crisis first, and then address our military needs."[41] These comments demonstrate that he either ignored or did not believe Manson's comments about the massive job creation in the computing and engineering sectors throughout Canada that the NSA/NSH would bring. He also ignored the compounding cost of replacing the helicopters later if, indeed, he was serious about the commitment to them. Practically, Chrétien could have solved two of the nation's "needs" at once. The Liberal leadership, however, refused to accept the use of military industry to create jobs and economic growth. Chrétien knew that it was far easier to sell education and social programs in Canada and that he could gain valuable political advantage if he portrayed the Conservatives as wasting the hard-earned money of taxpayers on the tools of war in a time of recession and peace. In short, he knew Canadian sensibilities well and how to spin expenditures on defence.

One of the organizations opposed to the helicopter replacements was the Canadian Peace Alliance. In a document asking for financial assistance from its supporters, it wrote that "there is simply no justification for this purchase – on any grounds. It must be stopped ... The government is on very shaky ground and they know it."[42] The document went on to discuss the government's "insatiable desire for the latest in modern attack and war-fighting weapons." Regarding the NSH portion, it claimed that "the entire fleet of these search and rescue helicopters could be upgraded *to top capability for just 1/20th the cost of the new helicopters* – leaving $4 Billion available for other programs" (emphasis added). The organization was clearly using misconception and outright fallacy to accrue funds for an anti-helicopter campaign. In one instance, these funds were used for a $10,000 full-page advertisement in *Maclean's* magazine in September 1992 attacking the program as "an obscene waste of tax dollars." Both sides of the helicopter debate, therefore, were investing a great deal of time and money in lobbying Canadians about the NSA/NSH.

In January 1993, Chrétien made the bold move of claiming publicly that he would cancel the project if elected. The Liberals had previously stated that they would review it before any action was taken. In a speech to the Faculty of Law at the University of Ottawa, he articulated that "the firm objective of a Liberal government over its term in office would be to reduce the deficit in absolute terms and as a percentage of Gross Domestic Product." He continued:

> And in no area do we differ more from the Conservatives than in setting priorities. For example, and there are many examples, a few months ago, the government committed $4.4 Billion ... over twelve years to the purchase of military helicopters to replace the Sea King fleet. $4.4 Billion ... One of the biggest single government expenditures in the history of Canada ... With the end of the Cold War, I consider such a lavish expenditure on anti-submarine equipment to be irresponsible. I am serving notice today that a Liberal government would seek to cancel this extravagant expenditure.[43]

Chrétien concluded by stating that the Liberal government would meet the national military needs "in a more realistic and cost effective way" and that "it must be understood that if any new spending programs are absolutely required, they must be directly related to the promotion of economic growth and jobs." The last statement sounded as if it had come from the NSA/NSH project office. From that point forward, Chrétien became the embodiment of the criticism of the helicopter replacement. And he did so with all the political acumen that had allowed him to rise to the position of Leader of the Opposition. Already his words were spun in a way to deceive the public. He began using the term "attack helicopters" to create the impression that they were only to be used to engage an enemy and draw Canada into war. Canada did not have, nor did it ever seek to purchase, this capability. The central roles of ASW and SAR were omitted from subsequent speeches by the Liberal leader.

The Conservative reaction to the speech was mostly one of uncertainty. They thought that Chrétien might be testing the program as an election issue. The Liberals held a commanding lead in the polls at the time. But a briefing note on the remarks stated that, "in Canada (as elsewhere), politicians have an abysmally poor record in keeping election promises. In this case, given the profoundly adverse impact of cancellation on Canadian industry, the economy and many members of the voting public, as well as the horrendous cost of termination, the speech should not be taken as the last word on the subject."[44] Regardless of what Chrétien intended by the comments, he had effectively made the NSA/NSH project a concrete election issue and would link the acquisition of new helicopters to the inability of the Conservative government to solve the deficit problem. He was determined to demonstrate that his party could succeed where the Mulroney Conservatives had failed. By February, the opposition attacks on the NSA became more concentrated, and the issue was commonly debated in the House of Commons. Chrétien began referring to the helicopters as "Cadillacs" and stated that the program was "absolutely unnecessary" and that unemployment was the priority.[45] Whereas sailors used to call the *St. Laurent* class Destroyer Escorts (DEs) "Cadillacs" because they were the best and most modern of their kind, when applied to the *EH-101* the term was meant to imply that they were ostentatious and unnecessary. Most of the Liberal ire was focused on Minister of National Defence Kim Campbell. Although she maintained that the military capability was necessary, she also defended the purchase as a way of creating highly technical jobs and stimulating Canadian industry.[46]

The other major tactic used by Chrétien was to spread the idea that ASW helicopters were useless because submarines were no longer a threat. Outside the House of Commons, Fred Crickard of Dalhousie University exclaimed that "I can't believe they are that ignorant. They're just trying to score cheap political points. It's an extremely cynical approach."[47] Martin Shadwick of York University echoed the sentiment and explained that the helicopter was being purchased for its multi-purpose capability: "I assume that as a sovereign nation, we still want to know what's going on within our 200-mile limit and the sea approaches to our territory." Indeed, the policy of maintaining a balanced maritime force to ensure effective surveillance and control of Canadian sovereign waters and maritime approaches was stated in the 1987 defence policy and reiterated in the April 1992 defence statement.

As part of their attack on the Conservatives, the opposition parties claimed that other types of new helicopters should be considered for the contract since they were cheaper. They included the Sikorsky *H-60* and *S-92*. These politicians also wanted to include Aerospatiale's *Super Puma*. They obviously ignored, or more likely were unaware, that an exhaustive process had already been undertaken to do just that; EHI had been the winner by a landslide. Sikorsky had already admitted that its available models did not adhere to the RFP.[48] Even by 1993, it was "generally recognized that the performance of the EH-101 is in many regards superior to that of all the alternative helicopters."[49] The model later put forth by Sikorsky, the *S-92,* was, as already mentioned, a "paper aircraft"; it was still on the drawing board and innumerable years away from certification. Furthermore, the higher level of capability of the *EH-101* justified the moderately higher price tag.

In February 1993, there was a question and answer session by the Director General of Public Affairs to the NSA project office on the increasingly precarious procurement of the *EH-101.*[50] The Director General inquired why the NSA required an ASW capability when the threat that it had been designed for no longer existed in the post-Cold War era. The NSA project manager's response stressed that there were approximately 900 submarines in the world at that time, owned by about 44 nations. Of those, 565 belonged to non-NATO countries, including, among others, 245 in the Commonwealth of Independent States, 122 in China, 44 in Latin America, 21 in North Korea, 6 in Libya, and 3 highly modern subs owned by Iran. And these nations were not halting their submarine building and procurement programs. The argument, therefore, was that there was still a modern and credible submarine capability that could be used against Canadian, NATO, or UN interests and that the *EH-101* was the best option to evaluate this situation and multiply the effectiveness of the new frigates and modernized destroyers. After all, Canada had the longest coastline in the world and was bounded on three sides by vast oceans. Moreover, submarine surveillance was not just a warfare situation; there were also passive requirements of verification and supervision of submarine activity if Canada chose to enter a theatre where there were submarines. Even more prescient was the statement that, "although the aggressive intentions of the Soviet Union have disappeared, that does not mean the potential for conflict and crisis has disappeared as well. The recent Persian Gulf War, its still ongoing aftermath, and the current strife in the former state of Yugoslavia bear testimony to the instability in the post Cold-War situation." Few people had predicted the rapid collapse of the Soviet Union. It was equally difficult to foresee what new threats might emerge. The tandem of the new frigates and the *EH-101* was meant to provide the flexibility and multi-purpose naval system necessary to adapt quickly to new security threats.

Another question that was posed was simple: "Why are we buying the EH-101, which is the most expensive of the available options?"[51] The answer was equally simple: "The EH-101 has been selected because it is the only helicopter that meets our shipborne and search and rescue requirements in a cost effective fashion." The representatives of the NSA/NSH project explained further the advantages that the *EH-101* had over the *Sea King*. It could operate in winter "icing conditions" on the Atlantic coast due to its de-icing capability. The *Sea King* or the *Labrador* could not. The new helicopter would also be able to remain in a hover if an engine failed as it had three engines instead of two. The increased range for coastal patrol and the increased cabin capacity for SAR, medical evacuation, emergency response transportation, and disaster relief were also stressed. The case of the *Sea King* in airlift support of peacekeeping troops in Somalia was used as an example where the *EH-101* was considered to be "4 times more effective in these operations." As for the need for the NSA to help with SAR, the project manager explained that the navy had used the *Sea King* to conduct these missions at sea beyond the range limits of the shore-based *Labradors* thirty-two times during 1990-91, which contributed to fifty-one lives being saved.

The next question was in regard to the size of the *EH-101* and the fact that the navy's ships would have to undergo modifications to be able to accommodate it. The NSA representative explained that the selection of any helicopter would require a number of ship modifications in the areas of aircraft handling, maintenance support, and special test equipment, and the twelve new CPFs needed work done only in those areas. Only the four 280 *Tribal* class destroyers required more extensive modifications, such as strengthening the flight deck and modifying the hangar to accommodate the new aircraft's size. The required changes for the twelve CPFs, four *Tribals*, and two AOR ships were estimated at $65 million. These costs were included in the NSA project budget and represented just over 1 percent of the total project costs.[52]

Perhaps the most pertinent question was this: "In these times of fiscal restraint, where is the money for this project coming from?"[53] The NSA team responded that the funds for the project had long been identified in DND budget estimates. In response to deep reductions in defence expenditures – nearly $6 billion since 1989 – the department had adopted a strategy that envisioned reductions in full-time personnel, infrastructure, and operating costs in order to focus on the purchase of essential equipment. Modernization of the equipment used by the maritime forces was included in these essential purchases. The strategy, the team explained, was to re-equip the CF with flexible and effective equipment that could perform a variety of tasks over the course of its service life. The *EH-101* easily fit within that strategy. As to the "ever present risk of cost overruns in defence procurement projects," the NSA team explained that they were often attributable to the design and development of new equipment and that this risk was being kept to a minimum in the NSA. The equipment being used was already in production. And instead of insisting on building the aircraft in Canada as originally planned and increasing production costs by roughly 50 percent, the NSA was taking advantage of a European production line already assembled. The contractors had also committed to a "firm, fixed price," and payments were tied to achievement of milestone events so that progress could be properly monitored and assessed. After a direct query about the cost of cancelling the project outright, the NSA team explained that it would likely cost taxpayers hundreds of millions of dollars but that the true loss would be measured in lack of job creation. "One of the biggest losses in canceling the project would be to the Canadian public – 45,000 person years of employment would be lost, in addition to the developments and progress which would occur in the Canadian aerospace industry." This loss was measured over the expected thirteen-year acquisition phase. The massive potential for IRBs was explained in full. It was also reiterated that the CPF program would lose an enormous amount of capability if the NSA program was cancelled, capability that included the potential for a CPF to defend itself against major advances in weapons technology that had resulted in long-range missiles that could be launched from either ships or submarines.

A major problem with the NSA project was that its management claimed that they were using equipment already in production; however, this claim applied only to the airframe being assembled in Europe. The mission suite to be used within it had been under discussion for almost six years, and much of it was new technology. Management had already admitted in 1987 that some of the software to be used in the mission system was under development and that there was a risk involved in "going where others have not tread." So, while it was true that EHI may have stayed within the budget decided on in 1992, the real threat to the NSA program was that the project had already required a great deal of time and money to that point.

Although the choice of the *EH-101* was not based on politics, the remainder of the procurement process was completely politically driven by IRBs and the necessity of including Canadian content in the mission suite. It was the Canadian government's policy for any major military procurement, and EHI had to comply. The Contract Definition Phase alone took five years. The entire acquisition of the new models after the final contract was signed was determined to take another thirteen. By the time cabinet gave its final approval to carry out the NSA/NSH program in 1992, the political climate in Canada had changed. The program had been under way for too long with little result. By 1993, the existing recession and the necessity of an election by November turned it into a giant political target. It was clear that the procurement officials had chosen the correct aircraft after meticulous analysis. As with the original selection of the *Sea King*, procurement officials decided on the most capable helicopter and then set out to acquire it. The amount of time and money invested in the project at NDHQ meant little to the opposition Liberals. The fact that there was a signed contract for the new helicopters was also irrelevant. They saw their opening.

A primary focus of the Liberal attacks on the Conservatives was the mismanagement of Canadian public funds and the national debt. The Liberals began to rally around reducing the national deficit and creating jobs. With this as their strategy, the multi-billion-dollar program to purchase helicopters for the military became an obvious object of criticism. The Liberals, led by Jean Chrétien, effectively attacked the project for its high cost, partly due to the combination of the NSA and NSH projects. But perhaps more significantly, they argued that these "Cadillac" helicopters were not needed in the first place. Party members expounded that the Cold War was over and that the submarine threat to Canada had dissipated with it. Chrétien wisely kept his assault narrowly focused by calling the *EH-101s* "attack helicopters" whose only function was to fight the last war; according to him, they served no other purpose. And so the NSA/NSH project became a central election issue and one of calculated political misinformation by the Liberals.

HMCS *Ville de Québec* exchanges its CH-124 *Sea King* helicopter for another on 31 October 2008. The helicopter being taken off the ship requires routine maintenance not normally carried out at sea.

Source: CF photo by Cpl. Dany Veillette. *Courtesy of National Defence. Reproduced with the permission of the Minister of Public Works and Government Services, 2009*

7
The 1993 NSA Cancellation: Money for Nothing

> The Sea King's basic range and endurance deficiencies cannot be resolved by any form of system update. Compounding the situation is the fact that the life expectancy of the Sea King has already been extended from 1985 to the year 2000, when it will be almost 40 years old. Further extensions beyond this point could be risky and would certainly be wasteful considering its very limited effectiveness.
>
> – Maritime Command and Maritime Air Group Headquarters, 29 January 1993

> I'll take one piece of paper, I'll take my pen, I will write zero helicopters, Chrétien. That will be it, and I will not lose one minute of sleep over it.
>
> – Prime Minister Jean Chrétien, 1993 Election Campaign

> The contract was terminated by Jean Chrétien within hours of being sworn in as Prime Minister. That decision, which has been compared with the cancellation of the Avro *Arrow* 35 years earlier, forced me to lay off more than 750 employees the following day, and it set the maritime helicopter procurement back by at least 15 years.
>
> – Former President of Paramax, Paul Manson

In February 1993, a memorandum on the costs of cancelling the *EH-101* contract was prepared by the research branch of the Library of Parliament. It stated that

> It is almost certain that the prime contractors and their subcontractors are now actively gearing up for production by accelerating the drafting of designs, ordering the fabrication of parts, and beginning the development of various equipment as would normally be done for any production contract. Cancelling the acquisition project now would stop this process in its tracks at a very early stage, but there would still be significant costs involved.[1]

The memo went on to explain that the longer the project went on the more work would be completed and the greater the costs of cancellation. It speculated that, because European Helicopter Industries (EHI) also had orders to fulfill for the British Royal Navy (RN) and the Italian navy, cancellation of fifty helicopters could increase the overall cost per aircraft for the other contracts. These increased production costs could be passed on to Canada. The memo also concluded that, since 83 percent of the electronic systems was to be supplied by Canadian firms, the other prime contractor, Paramax, had signed with several other subcontractors, such as Industrial Marine Products (IMP) in Halifax, which was to install the electronic system. Hence, "significant costs have probably already been incurred." Significant, and perhaps not surprising due to the consequences involved, was that there was no recent precedent in Canada "which would indicate what happens when a contract is cancelled." The memorandum noted that, "while a number of equipment purchases have been cancelled since April 1989, these involved proposed capital projects for which acquisition contracts had not been signed or planned phases of the acquisition process for which contracts had not been signed." Thus, there was no guideline to help determine how costs would be calculated in the event of cancellation. This only makes sense since there was enough time spent analyzing the contract beforehand to decide whether it was something that the government wanted to undertake. This is especially true of the *Sea King* replacement, which had been discussed since the 1970s. Finally, the memo brought up the obvious fact that, if this contract was cancelled, the government would likely have to start the process all over again; the costs to do so would have to be tallied with the costs of cancellation to determine how much replacing the helicopters would cost taxpayers in the end.

Part of the national debate on the New Shipborne Aircraft (NSA) centred on whether Canada could get by on a less capable model. Martin Shadwick testified in May 1993 that "the EH-101 is arguably more helicopter than we require ... The difficulty I keep running into is that the alternatives are arguably less than we require."[2] The Department of National Defence (DND) had already argued that it needed the superior performance of the *EH-101* to carry all the equipment necessary to fulfill complex missions, to have sufficient range to undertake extended surveillance and rescue operations, and to carry them out in all-weather conditions. These requirements applied to both the NSA and the New Search and Rescue Helicopter (NSH) versions, which had to operate in the North Atlantic and the Rocky Mountains respectively.[3] But clearly the issue raised by Shadwick – whether a robust Anti-Submarine Warfare (ASW) helicopter was needed in the post-Cold War era – was central to the debate on the *EH-101* acquisition. There were already questions about why Canada had been negotiating, since 1990, a $250 million procurement of *ERYX* anti-tank weapons with Aerospatiale in case of a massive Soviet land attack.[4] Those in the public who knew about the project believed that this threat no longer existed. The rhetoric was the same for the new helicopters. This was not for lack of trying by those involved in the NSA/NSH project. Paul Manson, President of Paramax, toured the nation personally to hold discussions with the editorial boards of all the major Canadian newspapers. He later wrote that "my objective has been to help people understand why the New Shipborne Aircraft/New Search and Rescue Helicopter project is a very important procurement for Canada, why the EH-101 is the only viable solution, and why the program should proceed now rather than later."[5] National Defence Headquarters (NDHQ) had even hired a public relations company to help sell the project to the people of Canada.[6] Notwithstanding a public information campaign, the advantages of having a flexible helicopter to adapt to any new security environment in a variety of roles were either not fully understood or ignored by the public. According to one former 423 Squadron Commanding Officer, this lack of knowledge was partly because "the military leadership had become complacent after the contract was signed."[7] Indeed, there was reason to believe that it was a done deal. A signed contract with the Canadian government has traditionally meant just that.

By the time of the election of 1993, the NSA/NSH project was not the primary concern of the Canadian voter. Although Mulroney had retained a parliamentary majority in the 1988 election, by 1993 there was ubiquitous public resentment toward him over an economic recession, free trade, the proposed Goods and Services Tax (GST), a fracturing of his political coalition, and the inability to bring Quebec into the 1992 Constitution at Charlottetown. Mulroney announced his resignation as Progressive Conservative leader and Prime Minister of Canada in February 1993 and was replaced by then Minister of National Defence Kim Campbell in June.[8] But between February and June, Campbell was still the defence minister and was the frontline defence of the NSA/NSH program.

The attacks on her credibility came from everywhere, and the NSA was caught in the middle. One critic claimed that the new ships could not hold the weight of the new helicopters. Campbell answered correctly outside the

House of Commons that "the Canadian patrol frigates were designed from the keel up to hold a helicopter of the weight of the EH-101."[9] As Peter Charlton and Michael Whitby later explained,

> The design of the helicopter facilities for the Canadian Patrol Frigates was aimed at operation of a single EH101 helicopter, a machine that is considerably larger and heavier than the Sea King. Indal Technologies Inc. of Mississauga Ontario have, since the closure of Fairey Aviation, been responsible for the support and further development of haul down systems, now known as Recovery, Assist, Secure, and Traverse (RAST) system[s]. Accordingly, the company was selected to install the RAST systems for the CPFs [Canadian Patrol Frigates]. Using extensive computer modeling capabilities, INDAL proposed a number of improvements over the DDH280 systems but the overall concept of using a constant tension wire and Rapid Securing Device (formerly the Beartrap) was retained.[10]

It was also being stated in the press that the "downwash," or force of wind created by the rotor blades underneath the helicopter's hover, would drown the very people whom the crew were trying to save. This was repudiated at a Standing Committee on National Defence and Veterans Affairs (SCONDVA) hearing on the NSA/NSH program on 13 May 1993 by Minister of Supply and Services Paul Dyck; he explained that a Canadian Forces (CF) team had experimented with the downwash in 1991 and that it did "not pose a problem for rescue."[11] In fact, the first journalist to offer this misconception, Dale Grant, later admitted to the SCONDVA committee that, after further analysis, he was "wrong, and unreservedly," and "able to eat a bit of crow."[12] His subsequent statement was really the crux of the matter of misinformation on the NSA/NSH program in the media. He stated that members of the press "are not always accurate and I don't believe we're required to be accurate and if everything I reported that everybody told me, if I had to assume it was absolutely true, I don't think there'd be much in the newspapers."

Another common critique was that the program was too expensive during a time of recession. During the review of the NSA/NSH purchase by SCONDVA on 29 April 1993, Member of Parliament (MP) Bill Rompkey stated that perhaps it would be best to wait "until we're in a better fiscal position" to carry out the program. But as MP Robert Hicks pointed out, this argument was somewhat extraneous as it would not change the defence budget allocated by the government:

> If the helicopters were cancelled all of a sudden, I don't think the government would take away [the] $12 Billion allocation to National Defence. I think you [DND] would look around to the next item on your priority list to upgrade capital equipment which could be armoured personnel carriers,

> it could be a multitude of things. So I can't see personally where the government of Canadians would save any money by canceling the helicopters.[13]

He believed that the government should let the "professional people decide how to spend it [the defence budget]." Hicks then brought up the misconception that the entire $4.4 billion would be paid in one lump sum; he made it clear that this money would be spread out over approximately thirteen years. It was hypothesized that the DND would save only $200 or $300 million that year if the contract was cancelled and that the money would simply be diverted to another priority. Hicks correctly stated, "I think we've identified, among us here, one problem, that is, the genuine ignorance of the public on this entire project ... We didn't sell the program ... I don't think anyone thought it was necessary." Ray Sturgeon, the Assistant Deputy Minister (Materiel) (ADM Mat), responsible for ensuring effective materiel acquisition and logistics support for the CF and DND, explained that, although the DND had a "relatively good public affairs process," the fact sheets being released by the DND as responses to erroneous reporting were not making their way into the press. He explained that the media created a negative perception among the public that the DND was having trouble countering. Although many tried to refute the myths being circulated by the Liberal Party about the *EH-101* purchase, the subject had become so politicized that the facts no longer mattered.

Hicks then asked Chief of the Defence Staff Admiral John Anderson if Canada's North Atlantic Treaty Organization (NATO) allies thought that the NSA acquisition was necessary to fulfill its commitments. He responded that clearly the British and Italians agreed with the procurement since they had also placed orders to acquire the *EH-101;* these navies were leading the way regarding maritime forces within NATO. He also maintained that the capabilities of the helicopter would allow flexibility for whatever form the security environment would take in the future.

At the next SCONDVA hearing on 5 May 1993, Minister of Defence Kim Campbell made a presentation. She pointed out that the government had made further cuts to defence expenditures in the April budget and that they were part of a series that had begun in 1989. She explained that her department's planned spending for the period 1989-90 to 1997-98 had been reduced by approximately $14 billion. She revealed that many projects would have to be cut or deferred, such as the *CF-18* modernization. Campbell maintained, however, that the one project still deemed essential was the acquisition of the new fleet of helicopters. She pointed out that this capability was still vital for the Canadian Forces to carry out the defence policy. "Unless we are going to get out of the business of ship-borne helicopters, unless we are going to get out of the business of search and rescue, then the Labradors and the Sea Kings have to be replaced."[14]

Campbell then addressed the issue of cost and the misconceptions among the public. The helicopters themselves were valued at $1.6 billion, but by then the public was hearing $5.8 billion to account for inflation. Campbell correctly maintained that these figures were being misconstrued in the press and that it was not being made clear that the $4.4 billion would be spread out over thirteen years. The full project cost also included the project itself and training, maintenance, and operations for that entire period. She illustrated that, if one were to buy a $15,000 car that way, its stated cost would actually be $75,000. The Canadian people had never been made aware that the actual cost of the project for that budgetary year would be approximately $200 million and that it had been set aside long ago. But this explanation would not have been to the political advantage of the opposition; therefore, the figure of $5.8 billion spent that year continued to be put forward. As MP Stan Darling put it, "so, again, that's a misconception that the majority of the people, if you picked them out of the phonebook, just names, they would all say that they thought the whole damn amount of money was being not only committed this year, but spent this year." The Liberal tactic of calling the *EH-101s* "attack" or "assault" helicopters to be purchased at the expense of programs such as child care also persisted. The wisdom of spending that much money on helicopters without the Soviet submarine threat was also questioned endlessly. These strategies by the opposition were ubiquitous in their questions during the SCONDVA hearings in April and May 1993.

The professionals involved in the project stayed firm in their defence of it. Paul Manson, for example, considered by some as "the most knowledgeable of all the witnesses," repeated his commitment that the *EH-101* was "exactly the right airplane for Canada at this point."[15] Although he was obviously an interested party because of his involvement in Paramax, his experience with procurement in Canada could not be overlooked, and his observations echoed those of defence analysts across the country. In particular, Peter Haydon of the Canadian Institute of Strategic Studies pointed out that the *EH-101* acquisition remained "a necessary step" to fulfill the government's decision on multi-purpose combat capable forces taken "some 10 years earlier." Martin Shadwick also maintained that ASW was still a valid requirement in Canada for security and sovereignty and that the lack of a specific threat in a changing security environment was precisely the reason for a multi-purpose flexible helicopter capability.

Campbell's fight to save the NSA/NSH program only became more heated after she won the leadership of the Conservative Party in June 1993. The Liberals had entrenched their opposition of the helicopter replacement in their official election document. It was entitled *Creating Opportunity: The Liberal Plan for Canada,* more commonly referred to as "the Red Book." The first two priorities listed within their "Fiscal and Monetary Policy" were the cancellation of the *EH-101* purchase and the reduction of spending on national defence.[16] Also included in the list of promises was the removal of the GST. Despite the rhetoric that the cancellation would lead to more jobs overall in other sectors, early predictions noted the massive amount of work and investment that would be lost if the NSA/NSH project was cancelled. By September 1993, the Campbell government had realized that it needed to mitigate criticism of the program. At a time when social programs such as health and day care were suffering from fiscal restraint, the government had difficulty justifying the NSA/NSH program. Further controversy came with the news that a prototype of the *EH-101* had crashed in Europe and killed four people; the problem was investigated and solved shortly thereafter.[17] Polls showed that four out of five Canadians did not favour the purchase.[18] Campbell, therefore, directed the DND to examine options for reducing the costs involved. Essentially, she discussed the program with her defence minister, Tom Siddon, and they brought their questions to Colonel Ed Fairbairn, Director Aerospace Requirements Maritime and Rotary Wing. They asked frankly about the minimum number of helicopters they could buy. He eventually responded with the figure of twenty-eight. Colonel Laurence McWha later stated that, "at that point, I knew it was over."[19] Campbell subsequently announced that Canada would procure a reduced fleet of forty-three *EH-101* helicopters – twenty-eight NSA and fifteen NSH – which was seven fewer than the originally planned thirty-five NSA contracted for. She said that the reduction would cut $1 billion from the contract cost. It was accepted that operational flexibility would suffer and that some aircraft normally used for training could be transferred to an operational role if necessary.[20]

Even given the reduction to the program, the NSA/NSH remained a major political liability for the Conservatives, and Chrétien's Liberals continued to threaten cancellation. Chrétien acridly avowed that "we don't need them, we don't even need one. Forty-three is as ridiculous as fifty."[21] In fact, the reduction actually hurt Campbell's credibility. She had been adamant as Minister of National Defence that a small cut in the number of helicopters was "not a particularly effective way to save costs."[22] She had maintained that every helicopter was essential to the Canadian military. In an attempt to reconcile this inconsistency, only days before an election was expected to be called, Campbell told reporters after a luncheon speech on 3 September 1993 that "I am the Prime Minister now and I am responsible for the entire government and the entire question of how we are going to get to a zero deficit in five years."[23] At another press conference after the announcement, Minister of Defence Siddon claimed that the reduction was the best decision as he fielded hostile questions on the helicopter deal for roughly two hours. He finally became exasperated and stated, "sometimes you guys are so preoccupied with the truth, that you don't understand that time changes a lot of things."[24] Indeed, there was much that the critics did not understand. For one thing, the DND had stated that cutting even five of the helicopters would not amount to more than $150 million in savings. Suddenly, the number was $1 billion with the removal of seven. No detailed figures were produced, but generally it was posited that $650 million could be saved by removing the seven helicopters, their mission suites, associated logistic support (spares, training, manuals), and project management costs. A further $150 million were believed to be removed as fewer contingency provisions were thought to be necessary due to lower risk associated with the project. Finally, another $200 million in inflation costs were considered to be removed.[25] These figures were rejected by some as a bookkeeping manoeuvre since many of the costs were fixed and not contingent on final numbers.[26] Indeed, the relative cost of the helicopters coming at the end of the production line was significantly less than that at the beginning. EHI was also not consulted on the matter. Shadwick said, "I wish them luck, but I can't see it – I think they're using a different calculator than everyone else. I think they're being very optimistic."[27]

Campbell had also contradicted her military advisers, who had always maintained that fifty helicopters met the minimum Canadian requirement. The Chief of the Defence Staff had stated to SCONDVA the previous April that "there is no examination going on at the moment in terms of reduced numbers ... We believe that the 35 for the maritime application and the 15 for the search-and-rescue application fit the minimum needs."[28] The Liberals saw the retreat from this position as a sign of weakness and indecision by the Conservative leader. Ed Goldenberg, the principal secretary to Chrétien, stated enthusiastically that "she [Campbell] has done more to make this an election issue than we could ever have."[29] It was also postulated that the DND did not even need the thirty-five helicopters and that it had inflated the total just in case it had to cut the number down further. As one journalist wrote, "if this is true – and at least two other sources suggest that it is – DND's credibility, never high with the media and the public, goes right out the window."[30] Many also claimed that the Conservative Party's talk of deficit reduction was incompatible with a multi-billion-dollar deal to buy helicopters. It was also claimed that, despite polls that showed the unpopularity of the helicopter deal, the Conservatives could not turn their backs on it as it was a patronage procurement to secure votes from Quebec.[31] As one editorial expounded, defence contracts in Canada "are among the tastiest cuts in any government's pork barrel," and purchasing decisions were often made not on merit but on the jobs that they would create and the votes that they would garner for the government.[32] The author claimed that the government had already given a contract for 200 *Bison* armoured vehicles to bail out a struggling General Motors defence division in Canada "even though it had no staff to maintain and service them." Clearly, Campbell and the Conservatives had misplayed their hand in the politics of procurement, and the NSA/NSH program received even more negative political attention.

The Canadian companies involved in the NSA/NSH project had been carefully monitoring all these events and relied on political and electoral considerations at every phase of the acquisition. They knew how procurement worked in Canada, and they continued to lobby those in power to sustain the project. One document prepared by Paramax revealed that the company placed a high premium on its lobbying capability: "Interaction with Crown at political level essential to success."[33] A Paramax spokesman, John Paul Macdonald, correctly maintained that there was nothing unusual in the lobbying: "I think that's the case for any major crown project." It was also reported that the company had prepared contingency plans for a scale-back of the number of helicopters ordered over a year before the actual reduction. This was all done with the knowledge that an election was likely approaching and could jeopardize the whole process.

The weakness of the Conservative Party and the certainty of a coming election caused the NSA/NSH project manager, Harvey Nielsen, to prepare a "Transition Package in the Event of a Liberal Majority" on 18 October 1993.[34] ADM Mat Ray Sturgeon was also involved in the briefing. His position involved a major role in the planning and implementation of the Long-Term Capital Equipment Plan, and he had been deeply involved in the helicopter procurement to date. The memorandum attempted to explain to a new Liberal government the weaknesses inherent in any alternative to carrying through on the *EH-101* project. It explained that the new number of aircraft to be procured was considered the "minimum viable option" and that any attempt to simply upgrade the *Sea Kings* for safety and operational capability would not alleviate "their inherent limitations" due to age and their continued "limited effectiveness in their assigned roles." The focus on Industrial and Regional Benefits (IRBs) for the NSA/NSH project was also addressed again as the memo explained how they would be affected in the event of another upgrade over replacement: "The IRB's associated with the ... upgrade would be minimal since the system upgrade would not create the industrial spin-offs that the NSA/NSH project has."

Another briefing note for Minister of National Defence Tom Siddon on the consequences of an outright cancellation of the NSA/NSH contract was prepared the next day. Neilsen wrote that "a decision to cancel the NSA/NSH project would have financial, industrial and operational impacts."[35] He explained that the lost project investment at the end of fiscal year 1993-94 would be approximately $500 million and that the government would be liable for extensive termination costs because it would be electing to end the contract for convenience. He explained further that, since there was a signed contract, EHI and Paramax "may sue to recover their costs incurred during the definition phase." The Canadian aerospace and defence industry would also lose the $3.2 billion IRB package, which was considered to amount to 45,000 person years of work in Canada. Numerous companies would subsequently experience major cash flow problems as a result of lost business. Paramax, it was determined, "would have to downsize substantially."

Another briefing note written seven days later stated that

> It was also recognized that our new CPF [Canadian patrol frigates] and upgraded TRUMP [Tribal Class Upgrade and Modernization Project] ships' more modern capabilities would not compensate for the Sea King's shortcomings. The opposite is, in fact, the case. These ships were designed to work in a multi-threat environment with a modern multi-mission helicopter that would complement their capabilities, act as their primary surface and subsurface sensor and exchange tactical data on shared high speed computer-to-computer data links.[36]

Although the *Sea Kings* performed satisfactorily in minor roles, the lack of these abilities was clear in the Persian Gulf, Somalia, Yugoslavia, and Haiti, where *Sea Kings* had to be continually altered mid-operation to fulfill the new multi-role requirement. The Haitian and Yugoslavian missions were similar to those in the Persian Gulf and Somalia in that they took the *Sea King* out of its classic niche.

In March 1993, Her Majesty's Canadian Ship (HMCS) *Algonquin* was deployed to the Adriatic Sea in support of a United Nations (UN) blockade off the coast of the former Republic of Yugoslavia. Ethnic and political tensions in the region had erupted into violence, and *Algonquin*'s air detachment was tasked to search for and identify merchant shipping to determine cargo and destination. The goal was to prevent war supplies from reaching Yugoslavian ports. Colonel (retired) John Cody described how the mission in the Adriatic included "a submarine threat hidden amongst the complex waters and chaotic politics of that troubled part of the world. Hence, we [did] have a requirement to maintain our ASW skills."[37] Years of unrest also reached a climax in Haiti when a military dictatorship seized control from the elected president. Several nations joined the United States in imposing a naval blockade on the supply of petroleum and arms. HMCS *Preserver* and *Fraser* were ordered to Haitian waters and arrived on 18 October. Three *Sea Kings,* equipped with Forward Looking Infrared (FLIR) and a Global Positioning System (GPS), flew from Shearwater to Florida to embark on the ships. The helicopters monitored and carried out boarding operations on ships that passed through the blockade area to ensure compliance with UN resolutions.[38] It was later explained that, "during the Haitian Operation, all three Sea Kings have had to be replaced at sea with re-configured variants. This, combined with the Sea King's overall lack of capability and protection, will impose significant limitations in the use of ships in all operations."[39]

Canadians, dissatisfied with the performance of the economy and the Conservative government's management of the national debt, voted Liberal in the election of 25 October 1993. The cessation of work and the severance of ties with EHI were immediate following the election. Part of this cancellation involved stopping in-flight-icing trials for the *EH-101* scheduled to start on 1 November in Shearwater. In 1992, EHI representatives had visited a number of North American sites to determine the best location for such testing. This effort was completely separate from and unrelated to the NSA/NSH contract work. Shearwater was eventually selected primarily because the historical climate offered the best guarantee for the occurrence of the severe in-flight-icing conditions required for the trials. All services for the trials team, such as temporary office and shop spaces, electricity, and telephones at Shearwater, were arranged and paid for by EHI. This work had all been completed by the fall of 1993 in anticipation of the trials commencement date. The trials team was slated to arrive in Halifax on 29 October, and the aircraft was scheduled to arrive in Halifax by sea on 31 October. In the early evening of 29 October 1993, in his capacity as the Acting Wing Commander at Shearwater, Colonel (retired) Laurence McWha received a telephone call. It was the Executive Assistant (EA) to the Deputy Minister of Defence, Robert Fowler. He wanted to know who had authorized the *EH-101* flight trials and the use of Shearwater along with many other details. McWha explained that NDHQ had authorized the trials and that he was surprised the Deputy Minister was unaware of this. The EA then asked McWha to hold. As McWha recalled,

> I could hear him in the background repeating the information I had given him. I then heard the very loud and agitated voices of the DM [Deputy Minister] and VCDS [Vice Chief of the Defence Staff] as they gave instructions to the EA. I was able to hear their instructions very clearly. The EA repeated these to me when he returned to his phone. Under no circumstances was the EH-101 to be allowed to operate at Shearwater; under no circumstances was the EH-101 to be permitted to fly in Canadian Military airspace; and under no circumstances was the trials team to be allowed access to DND property.[40]

McWha then informed the EA that these "11th hour instructions" seemed to be entirely unreasonable but assured him that he would enforce them.

> I was not convinced that the DM had the right to unilaterally overturn international bilateral agreements and I was almost certain that his edict to

> disallow flight within Canadian Military airspace required Ministerial authority given that the various certificates and approvals for that had already been approved. It was evident to me that the DM's instructions were politically motivated given that the Liberals had been elected just a few days earlier on 25 October with one of several election promises being the cancellation of the NSA/NSH contracts. My assessment was that these unreasonable demands were being generated from within a political vacuum. Mr. Fowler's previous boss, MND [Minister of National Defence] Tom Siddon, had lost his seat. Siddon's successor would neither be named nor sworn in until 4 November. In my view, Mr. Fowler had no political bosses on 29 October 1993 and, unless he was receiving orders directly from the Chrétien transition team, was probably acting alone.

The *EH-101* arrived on schedule and was flown by the EHI crew directly to the Halifax International Airport, where it was stored in an IMP aerospace hangar while the crew waited for further instructions.

The NSA program and the *EH-101* contract were cancelled roughly six hours after Jean Chrétien and his government took power. It was the first major decision by the Prime Minister and his new cabinet. Chrétien maintained that it had to be done immediately to avoid further contract costs. There was no study of the repercussions of the cancellation and no indication of an alternative plan. It was stated, however, that there would be a defence review by the new Minister of National Defence, David Collenette. The Quebec Minister of Industry, Gerard Tremblay, claimed that it would cost companies in his province $1 billion.[41] None of the provinces would be compensated for work already completed or the massive layoffs that were imminent. "There is no compensation for anybody," Chrétien told reporters who asked about the expected job losses. "It might cost a lot of money to the taxpayers. It's very unfortunate. But I always made it clear, for a long time, that I would scrap this program."[42] Chrétien attempted to give the impression, therefore, that he would be a man of his word. He would show Canadian voters that he was a man of action and carry out the promises that he had made to them. After the first cabinet meeting, he repeated his oft-stated mantra that the Cadillac type of helicopter was no longer needed and that the Progressive Conservatives had ignored the new reality of the end of the Cold War.

The rhetoric on the end of the Cold War had proven successful for members of the Liberal government, and their leader could not end the NSA/NSH program fast enough. After the election, defence officials attempted to explain to Chrétien that immediate cancellation was not the only option available. ADM Mat Ray Sturgeon contacted the new government and attempted to explain that it could institute a "suspension of work" clause in the NSA/NSH contract and use the time to study the possible repercussions of total cancellation. Article F4 of the contract stated that

> The Minister [Public Works and Government Services Canada] may at any time, and from time to time by Notice, order a suspension of Work in whole or in part for a period of ninety (90) days following receipt of the notice by the Contractor. The Contractor shall comply with the order and promptly take all steps necessary to put the order into full force and effect, including suspension of all affecting subcontracts.[43]

Chrétien's argument that the procurement had to be cancelled immediately to avoid further cancellation costs was therefore inaccurate. The information provided to him by NDHQ was ignored. This demonstrated a high degree of obstinacy and an exceptional lack of foresight by the new government. Sturgeon has since called this action "outrageous."[44] Ironically, adherence to the election promise was never expected by many of those involved. As General Manson has proclaimed, "even after he [Chrétien] was elected, we still thought that there would be discussion with the military on the matter. We were wrong."[45] Lieutenant General (retired) George Macdonald has reiterated this belief: "We were still optimistic after the cancellation that Chrétien would come back and renegotiate. We had binders of information ready to go on why we needed them ... but we were never asked."[46] There had never been any evidence put forward by Canada's military experts showing that the strategic environment no longer required modern naval helicopters. The NSA program was cancelled without a credible alternative plan.

Canadian governments have largely failed to predict their military needs for future operations. As Canadian military historian Desmond Morton exclaimed immediately after the election of Chrétien in 1993, "our capacity to prophesy what we'll need our defence forces for, when, why and how, had proven so far to be zip."[47] Cancellation of the *Sea King* replacement is another example of this failure. As Morton explained further, defence policy in Canada is usually poisoned by partisan interests: "We decide what we're prepared to do in the way of defence, what insurance premium is appropriate. Then we decide whether we want it to be efficient or inefficient – and Canadians have generally preferred inefficient."

In Canada, as in most liberal democratic states, civil control of the military has meant control by civilians elected to Parliament acting in accordance with statutes passed by that legislative body.[48] Civilian control is intended to ensure that the decisions affecting the defence of a nation and the use of the armed forces are made by politicians who are responsible to the people – not by professional soldiers. It can also be argued, however, that the politicians have a responsibility to use the information that they are given by military experts, despite their possible bias, to make informed decisions. But they are under no obligation to do so. One author has stated that "civil control of the military is managed and maintained through the sharing of responsibility for control between civilian leaders and military officers."[49] Each side, therefore, agrees to assume certain responsibilities and accountabilities within a formalized regime of understandings. This regime should, theoretically, allow each a measure of independence, but ultimately the civilian authority reserves the right to make the final decisions – including those that affect weapons procurement.

In a time of recession, the NSA/NSH program became a symbol of misplaced government financial priorities. The Liberal opposition declared that the post-Cold War era was devoid of a submarine threat, and the Conservative defence was not able to counter effectively the misinformation campaign. The project was also highlighted by the fact that Kim Campbell was not only the defence minister in early 1993 but also the frontrunner in the race for the leadership of the Conservative Party after Mulroney stepped down. A federal election was imminent, and opposition parties had to select their focal points of attack. These two variables highlighted the helicopter replacement program to a degree that would likely not have existed a year earlier or a year later. As former Chief of the Defence Staff General Paul Manson stated to SCONDVA on 13 May 1993, "as an accident of timing, the program has been caught up in a whirlwind of controversy that is clearly associated with the political calendar in Canada this year."[50] To make matters worse, after Campbell was selected as leader of the party, the Conservatives consistently fumbled the NSA/NSH issue by being inconsistent and seemingly dishonest. These errors compounded the most debilitating problem of all – the lack of understanding by the public of the roles of the helicopter and the cost of the program in actual dollars spent during the recession of 1993.

Defence officials were consistently overpowered within the civil-military relationship in Canada regarding replacement of the *Sea King*. Procurement of weapons and equipment for the CF was controlled by political considerations over the advice of military professionals. Cancellation of the NSA program is a powerful example of the hierarchy between the military and the civil power in Canada with regard to equipment procurement; the contract was subjected to severe political scrutiny and interference outside the Department of National Defence. Although there had been much activity and significant effort devoted to replacing the *Sea King* by defence officials in the 1970s, 1980s, and early 1990s, they were essentially overruled in one day by their political leaders. Cancellation of the NSA limited the capabilities of the CF and suppressed the helicopter-carrying destroyer (DDH) idea that Canada had designed. The example of the *Sea King* has highlighted the civil-military conflict in Canada regarding armament procurement and illustrated that national defence contracts are often controlled by political considerations rather than military requirements. The pervasive fear of Soviet expansionism in the 1960s had convinced Canadians that a naval ASW helicopter was a necessity. The government, therefore, had the support of the people and was free to authorize a major procurement of helicopters as a result. By the 1990s, with this perception of threat removed, it became much more difficult to sell investment in ASW defence. The public had to be educated on the instability of the post-Cold War period and the utility and roles that the helicopters had acquired throughout their operational lives that continued to make them necessary. The Conservative government and the Department of National Defence failed in this endeavour.

CH-149 *Cormorant* Helicopters on tarmac at Sydney Airport on 12 March 2009.

Source: CF photo by Pvt. Vicky Lefrançois. *Courtesy of National Defence. Reproduced with the permission of the Minister of Public Works and Government Services, 2009*

8
The 1994 White Paper and the New Statement of Requirement: The Ghost of Procurements Past

> Air Command (AIRCOM) and Maritime Command (MARCOM) operational readiness and effectiveness reports have consistently described the Sea King as being operationally obsolescent due primarily to its outdated technology and limited sensor suite ... [The] Sea King has become a known maintenance burden as maintainability and reliability are suffering due to the age of the aircraft and its systems. Without a modern, reliable and maintainable aircraft capable of the full range of operations required in today's maritime environment, MARCOM will be unable to perform its assigned missions without the acceptance of substantial risk.
>
> – "Statement of Capability Deficiency," Maritime Helicopter Project, June 1994

After cancellation of the New Shipborne Aircraft (NSA)/New Search and Rescue Helicopter (NSH) project, there was protest in Quebec and Nova Scotia as both provinces stood to lose a large percentage of the Industrial and Regional Benefits (IRBs) from the NSA portion. After all, one of the primary contractors, Paramax, was stationed in Montreal, and Industrial Marine Products (IMP) was stationed in Halifax. Some writers, such as Lise Bisonette of *Le Devoir,* however, claimed that it was not until after Chrétien won the election that the protest began in earnest in Quebec. She pointed to the fact that Chrétien had first opposed the contract in the House of Commons on 9 June 1992 and, despite being "irresponsible" in the matter, had been completely clear from that point on his position. "Not one mayor, not one minister, not one chamber of commerce president told Mr. Chrétien, who had every chance of forming the next government of Canada, that his promise to cancel created any kind of problem."[1] Bisonette claimed that this meant one of two things: "Either all these neo-critics really have just awakened after 16 months of indifference to what is a major situation, or they deliberately withheld their fire to allow Mr. Chrétien to win." She thought the second possibility far more likely and that Quebec would now have to pay the price for its indifference to questions of defence:

> The government of Quebec, the Bloc Québécois, the unions and industrial associations are reaping the fruit of an irrational and collective indifference and hostility towards anything that touches defence. Whether they are "Cadillacs" beyond price – that is another question – they are not bazookas with propellers destined to kill little children on the orders of the powerful in some rebel territory. They are part of considerations touching on maritime sovereignty, security of boundaries, international cooperation whose obligations have not evaporated with the end of the Cold War. An independent Quebec would have to consider them as much as Canada.

Bisonette concluded by writing that Lucien Bouchard vilified the military industry "like an adolescent at his first demonstration." Michel Audet wrote that the protests by the Quebec and Nova Scotia governments were legitimate, but he correctly concluded that "they should have been expressed more firmly during the election campaign when Paramax was expressing its great concern." And where, as Don Macpherson wrote, was Premier Bourassa, "with his supposed obsession with jobs, especially high tech ones?"[2] The author even speculated that maybe Bourassa hoped for a quick transfer of control over manpower training in Quebec in return for his silence on the NSA/NSH project during the campaign. Whatever the reasons, Quebec leaders had made little or no attempt to halt cancellation of the program until after it was done.

The decision to cancel the project was embraced in some circles. Pam Frache, co-ordinator of the Toronto-based Canadian Peace Alliance, exclaimed that "it's absolutely fantastic news. And now we would like the Liberal government to move as quickly as possible on its other promises to convert defence industries to civilian production."[3] One editorial agreed that, since the Cold War was over, there was simply no need for the capability. The author went on to use the spurious logic that, even if the cancellation costs amounted to $800 million, that was less than what it would cost to buy the helicopters. The article ended by stating that Paramax should "step forward with courage and imagination" to seize its new challenge.[4]

In Montreal, Paramax employees stayed after work to watch Chrétien deliver his cancellation message on the cafeteria television. Mario Sabourin, one of several hundred engineers and technicians who were designing the computer and electronic systems, asserted that "it's a sad story, and the saddest thing is it's just politics."[5] Paramax had to fire 750 employees the day after the cancellation.[6] Buzz Nixon, a former Deputy Minister of National

Defence, also dismissed the cancellation as a major miscalculation: "When he [Chrétien] finishes the review and takes a look at the ups and downs of repairing the Sea Kings, he'll find that the only thing that makes sense is to buy these helicopters."[7] Although figures had not been decided on yet, it was also clear that the government was going to have to pay massive cancellation costs to the companies involved in the program.

Military procurement in Canada drastically changed after the NSA/NSH cancellation. In February 1994, Minister of National Defence David Collenette wrote a memorandum on the impact of the new Liberal budget on national defence. It explained that the "government will reduce expenditures devoted to National Defence by $7 Billion over five years."[8] This represented a 12 percent reduction in planned defence expenditures. And although the money from the *EH-101* project would be funnelled back into the Department of National Defence (DND) long-term budget, Collenette affirmed that the "department will still have to find money for replacement helicopters should the requirement be confirmed in the defence policy review." Perhaps the most pertinent section of the document was on the new acquisition strategy of the DND. It had decided to "emphasize the purchase of equipment 'off the shelf,' the use of commercial standard technologies, and unless ... absolutely necessary ... the avoidance of military modifications." It was concluded that the management personnel involved on the procurement side of the DND could be significantly reduced as a result.

In July 1994, a Special Joint Committee of Canada's Defence Policy was conducted, and questions were posed on the operations of the *Sea King*. The first question regarded how many hours of maintenance the aircraft required for each hour in the air, and it was revealed to be approximately twenty-five, up from nineteen a decade earlier. Clearly, the *Sea King* had always been a high-maintenance aircraft. More significant was the number of accidents and forced landings since 1984. In February 1993, a *Sea King* conducting a surveillance mission at night ditched into the Gulf of Mexico after suffering an electrical system failure. After one of the flotation bags also failed to inflate, the aircraft sank. The crew evacuated, and they were subsequently rescued by Her Majesty's Canadian Ship (HMCS) *Nipigon*. The most recent incident at the time was on 28 April 1994. A *Sea King* had caught fire in flight as a result of a ruptured fuel line and was forced to make an emergency landing outside St. John, New Brunswick. Although the landing saved two crewmen, Major Bob Henderson and Major Wally Sweetman died in the cockpit as the fire quickly engulfed the aircraft. Henderson, a father of three, had been interviewed after Chrétien was elected in 1993 and was quoted as saying, "we're not yet pessimistic. We're all hoping that in the sober light of day the Liberals will yet change their minds [regarding the NSA cancellation]."[9]

The non-fatal forced landings due to technical difficulties are perhaps more indicative of the diminishing air worthiness of the *Sea Kings* by 1994. There had been 118 such cases from 1984 to 1994 resulting from a range of problems from fuel leaks, main gear box oil pressure fluctuations, transmission oil pressure fluctuations, engine failures, and power reductions.[10] There was also the issue of whether there would be enough maintenance and spare parts to keep the *Sea Kings* able to conduct robust operations. As Colonel (retired) John Cody has maintained,

> As for the spare parts issue, it stands to reason that with a contract (about to be) signed, that it would be fiscally prudent to commence a run-down of available Sea King spares, ensuring you had enough to handle the expected introduction of the new aircraft and the carefully controlled phase out of the Sea King. So, as the planning was proceeding for the introduction of the new aircraft, staff at NDHQ were keeping a handle on the Sea King spares in inventory. When the government of the day cancelled the contract for the EH 101 it left us with insufficient spares for a long haul the length of which was anybody's guess, with the entire fleet of Sea Kings. This then caused a ripple down effect as NDHQ then had to commence a ramp-up of spares procurement for the Sea King. The Air Force had not planned for this and so it was a lengthy process getting money approved and letting contracts again to refurbish our parts bins at Shearwater. With long lead items at stake in some cases, we really had to be very careful and manage each part almost by name.[11]

The year after the *Sea King* replacement was cancelled a 1994 Defence White Paper acknowledged that

> There is an urgent need for robust and capable new shipborne helicopters. The Sea Kings are rapidly approaching the end of their operational life. Work will, therefore, begin immediately to identify options and plans to put into service new affordable replacement helicopters by the end of the decade ... The Labrador search and rescue helicopters will be replaced as soon as possible ... This role may be performed using the same helicopter that we acquire for the maritime role.[12]

The White Paper reiterated the policy of having a multi-role, combat capable force. But it also stated a variety of other domestic concerns to be added to the mandate of the DND, such as assistance in protection of the fisheries, drug interdiction, environmental protection, humanitarian and disaster relief, and potential demands for aid to the civil power. Helicopters clearly had a responsibility to fulfill in these areas. As a result of the continued recognition that the *Sea King* needed to be replaced, therefore, the DND went, once again, through the onerous process of creating a new Statement of Requirement (SOR) based on the White Paper. The Maritime Helicopter Project (MHP) was registered in the Defence Capital Program in April 1994; the first draft was sixty-five pages and was submitted to the appropriate staff officers on 25 October 1994.[13]

On 7 February 1995, the Commander at Maritime Command Headquarters, Rear Admiral G.L. Garnett, wrote a memorandum on the new SOR for a maritime helicopter:

> At the broadest level, the SOR cannot put the Department or government in an overly confining situation. This was the type of situation that developed over the EH-101 in that the rationale for the helicopter was based on classic open ocean war-fighting roles ... While ASW [anti-submarine warfare] and ASUW [anti-surface warfare] are primary missions for any helicopter operating from naval ships, they are no easier to sell to Canadians today than in the case of the EH-101. We need to present a full range of missions for the helicopter, including SAR [search and rescue], coastal surveillance, and peacekeeping/embargo operations.[14]

In fact, this concept was not a new one at all. The NSA project office had always stressed the multiple roles that the *EH-101* would perform beyond ASW.[15] The Liberals' justification for the NSA cancellation – that the *EH-101* was a Cold War aircraft only capable of ASW – was false. The NSA was required to be capable of ASUW, ASW, surveillance, and SAR/utility without reconfiguration. The helicopter was to be multi-mission capable and was never just about hunting Russian submarines. This reality had little impact on how the Canadian public felt about it. It was an expensive defence contract during a recession, and the people did not support it because of apathy, ignorance, or a belief that it was simply unnecessary in the post-Cold War era. The fact that much of the Canadian public had rejected the NSA throughout the election of 1993 made the creation of the new SOR extremely sensitive politically. Every move had to be carefully thought out. It took six drafts over an entire year just to get started. Admiral Garnett wrote that "I am reminded that the MND [Minister of National Defence] and, indeed, the PM [Prime Minister] have noted that we need a new maritime helicopter with Chevrolet like characteristics to complement our new patrol frigates."[16] In 1994, the Canadian Patrol Frigates (CPF) began to enter operational service. These highly capable ships, originally designed to operate in conjunction with the NSA, replaced the DDHs of the *Improved St. Laurent* and *Annapolis* classes. They would operate with the *Sea Kings*.[17] Cancellation of the NSA in 1993 ensured that the next attempt to replace the *Sea King* would be mired in delays and suspicions of interference. And it was clear that

Garnett was being pressured to deliver a SOR that would not fit a "Cadillac" like the *EH-101*. The existing beliefs of elected politicians were having direct influence on the professional recommendations made by those in uniform. The final result, therefore, was that the politicization of procurement was self-perpetuating.

During further discussions, Lieutenant Colonel J.F. Cottingham tried to ease his Commander's concerns regarding the uncompromising nature of the draft SOR; pointing to the fact that the SOR had been written in a way that "it can be changed at any time" and that it was "merely guidance from the operational community to the NDHQ [National Defence Headquarters] procurement staffs."[18] He also addressed the Commander's apprehension that the specifications in the draft SOR were too demanding:

> It must also be understood that all of the performance specifications were derived using the present Sea King as a baseline. In other words, *all* of the aircraft performance specifications called for in this document can be met by a number of helicopters including the SH 60/S7, Super Puma, S92, a Sea King (new or rebuilt) and some of the Russian made machines."

The *EH-101* was noticeably absent from the list. Cancellation of the NSA, therefore, had created a reduction in capability for a naval helicopter. The aircraft that had been dismissed as incapable of fulfilling the maritime helicopter role in the 1980s were suddenly deemed possible candidates in 1995. But the Commander still thought that the SOR was too demanding and that it was diverting his team toward the most expensive options; this situation had been a political liability for the NSA. Cottingham pointed out, however, that they were avoiding any new technological development this time and would attempt to buy off the shelf wherever possible, thereby reducing design and manufacture costs. The plan was obviously to stay away from the complications of extensive IRBs that involved emerging technology. But there would still be, he pointed out, the cost of purchasing "the long term support requirements up front." Cottingham then revealed a tone of trepidation in the letter:

> Overall, my impression of the SOR is favourable. It is not written or even structured in the way that I would do it but it is good enough to meet the needs of the project. As DARMR [Director Aerospace Requirements Maritime and Rotary Wing] stated ... timing is getting very tight for the issuance of a SOI [Solicitation of Interest]. I am concerned that if we spend too much time attempting to make the SOR perfect, we will miss the gate for Cabinet approval or worse ... be directed to buy a helicopter which cannot meet any of our needs or will cost us too much to operate in the long term.

Clearly, he understood the importance of timing in the Canadian procurement process and that it should not take excessively long to devise a SOR.

On 9 February 1995, Admiral Garnett sent back the SOR again. He still insisted that they had to stress the non-military roles, a more flexible platform, and "a more generalized statement of range/endurance not tied to war fighting capabilities."[19] Major A.D. Blair of the air force, later a Commanding Officer (CO) of 423 Squadron, responded by writing that he was very uncomfortable with these continued alterations since he thought that "what is being proposed is no less than a rewrite of current CF [Canadian Forces] policy to present a more publicly palatable procurement strategy. It may also ensure that the Bell 214ST fills the bill. This surely is a battle that must be fought at the highest levels and not one which should compromise a Statement of *Operational* Requirement." Blair concluded that the navy had lost sight of what a SOR was and that it "should not be a medium for policy change, or political correctness for that matter, but rather a true and accurate reflection of what the Maritime Community requires in order to accomplish its assigned missions." Indeed, it is unfortunate that complete threat assessments were not considered sufficient justification for the procurement. Even the Rear Admiral, Bruce Johnston, the Maritime Commander Pacific (MARPAC), wanted to pare down the requirements. He asserted to Garnett that "my first impression was that the MHSOR [Maritime Helicopter Statement of Requirement] was too all encompassing and uncompromising and that it could potentially lead us down the same road as NSA; namely to the unaffordable 'Cadillac' solution ... We cannot afford to be hung up on professional judgment criteria like performance and range for example."[20] The Maritime Commander Atlantic (MARLANT), however, did not agree with his colleagues and was extremely disappointed with how the process was going.

> After reviewing the ref A comments, I can no longer contain my sincere disappointment at the level and depth of the Navy's concerns. I don't think they fully understand the nature of an SOR as a pure, policy-driven statement; they are confusing procurement strategy matters and political perceptions with our mandated roles and desired capabilities. I find it passing curious that the Navy's east coast warfighting commander does not want to use a valid threat assessment as the basis of procurement for a weapon system – what should we use then?[21]

But after what happened to the NSA program, Garnett and Johnston definitely had a point; they believed that the navy might have to take what it could get. As Stuart Soward wrote in 1995, "the aircraft are now over thirty years old, and have more seniority than most of the pilots flying them."[22]

The SOR was jointly approved by the navy and air force on 13 March 1995 for the Maritime Helicopter Project. Part of the delay in getting this far was that both the navy and the air force had to agree. After unification ended, the air force essentially retained control of naval aviation procurement. Soward wrote that "Canada is the only country to allow another service to be responsible for their aviation, Navy and Army. They, therefore, control helicopter procurement to this day."[23] Notwithstanding the delays, the final document reduced the range and endurance requirements from the original NSA SOR by 25 percent. The ability to fly in icing conditions, for example, was reduced from moderate to light. It contained a far more general tone than that for the NSA, with no particular emphasis on any one maritime mission. It was also not as technically specific, thereby affording greater latitude to explore a variety of options.[24] Although the SOR pointed out that the emphasis on Canadian naval operations shifted from subsurface to more surface surveillance at the end of the Cold War, it correctly concluded that subsurface surveillance was still a vital capability in a period of continued submarine proliferation. This was especially true if the navy wanted to remain consistent with the multi-role capability. And the roles of shipborne helicopters were ever expanding. For example, there were more littoral operations in shallow coastal regions. The necessity of both surface and subsurface surveillance had been apparent in the Canadian monitoring of shipping in support of the naval blockade in the Adriatic, where the presence of five submarines from the Republic of Yugoslavia had made this capability essential to the safety of the ships involved.

The SOR, therefore, divided the maritime helicopter mission into three components: surface surveillance; subsurface surveillance; and utility operations. The number of helicopters needed to satisfy the MHP was established at thirty-two. A document entitled "Details of the Operational Requirement" explained that, for reasons of affordability, "no allowance was made for attrition."[25] This meant that it was considered the minimum amount needed to fulfill maritime missions, and any loss of aircraft due to accidents would limit operations. The document pointed out that the previous reduction of the NSA to twenty-eight by Kim Campbell "would have significantly reduced operational flexibility. In addition, this reduction would have resulted in compromised training, over worked crews, and ships going to sea without embarked helicopters."

A definable similarity of the NSA and the MHP was that there would be a public relations campaign to convince Canadians of the necessity of replacing the *Sea King*. Although such a campaign had been recognized as important for the NSA program, and many actors involved in the project had attempted to exert a powerful lobby for it, the strategies for the MHP would be more extensive. A "Communications Plan Activation" document explained that "following the announcement of the MHP it will be the responsibility of DDPA [Deputy Director Public Affairs] (Mat) to activate the MHP communications plan."[26] The major components outlined in the plan were as follows.

- *A national news release* would be prepared in conjunction with the Minister of National Defence's announcement.
- *Backgrounders* would be prepared explaining details of the MHP requirement.
- *Fact sheets* would outline characteristics of the maritime helicopter contenders compared with the *Sea King.*
- *Questions and answers* would be prepared to assist authorized spokespersons.
- *Speaking engagement notes would be prepared* for senior officers and other DND officials called on to speak on the matter.
- *Point-form information sheets* should be prepared to encourage speedy responses to letters to the editor.
- *Stock photos* showing the contending helicopters and the *Sea King* operating from ships would be made available.
- *Stock videos* showing the contending helicopters would be made available.
- *Third-party endorsements* would be sought from opinion leaders and subject experts, who would be briefed and provided with information packages following announcement of the MHP.
- *Media monitoring* of print, television, and radio would be undertaken at local and provincial levels by public affairs personnel at the commands and at regional DND public affairs offices.
- *Media trips* would be arranged to Shearwater and Pat Bay to acquaint media representatives with maritime helicopter operations and the importance of replacing the *Sea King.*

Although this was clearly a wise course of action considering the NSA experience, these plans had to wait until the project was officially approved and under way.

As plans to restart the process of replacing the *Sea Kings* were proving difficult, negotiations regarding the final costs of cancelling the NSA contract were also coming to a close. It was not until October 1995 that an agreement was reached with European Helicopter Industries (EHI) over the cancellation. On 23 January 1996, details of the agreement were announced jointly by David Dingwall, Minister of Public Works and Government Services Canada (PWGSC), and Enrico Striano, Managing Director of EHI. The minister stated that "negotiating this settlement has been a long process and I am pleased to say that the Crown and E.H. Industries have reached this mutually satisfactory agreement. It is fair to say that the Government of Canada has closed the books on the EH-101 helicopter program."[27] Specifically, the total costs of termination amounted to $478.3 million. The new agreement gave EHI $136.6 million for costs of work completed and work in progress at the time of termination, with an additional $21.2 million paid to the company for the outright cost of contract termination. The remaining $165.9 million went to the Loral company (formerly Paramax), the project's other prime contractor: $98.4 million for costs of work completed and work in progress and $67.5 million for cost of contract termination. There had already been $154.6 million paid out by the government before the cancellation to EHI and Paramax for the Project Definition Phase, research and development, and project implementation, bringing the total to $478.3 million. Although Chrétien had accused the Conservative government of Brian Mulroney of squandering taxpayers' money, the new Liberal government nullified years of work and investment and paid a fee of close to $500 million to cancel the contract. Although the final settlement in 1995 was expected to be far worse, this was still a substantial amount to pay since no actual product had been acquired for the money. In fact, the government had been prepared to pay far more for the cancellation. It had allocated $250 million in the 1994-95 Main Estimates for final termination costs, of which only $89.2 million had been required.[28] As Peter Haydon of Dalhousie University proclaimed, "it's way lower than any figure I've heard. But, then, business is business."[29]

The figure of $478.3 million is considered an illusion by some. Former Assistant Deputy Minister (Materiel) (ADM Mat) Ray Sturgeon has written "that [the $478.3 million] did not include the research and development investment that was made for next generation technologies in avionics and some mission capability. The Canadian taxpayer and the Defence Department made a quarter billion dollar investment in the defence industry and could not capitalize on it because of the cancellation of the EH 101 contract."[30] Indeed, the amount of work in the form of subcontracts to be fulfilled in Canada was substantial. There were numerous agreements between Canadian companies and EHI to provide parts and systems for the *EH-101* global market, which would have provided exceptional stimulation of the Canadian aerospace industry. Canadian industry had been guaranteed work for 170 shipsets of *EH-101* parts that it was supplying to the NSA. Although over two-thirds of the economic capability and activity in the aerospace industry was located in Ontario and Quebec, a number of large subcontracts outside eastern Canada would have stimulated economic development across the country. The contract that went to General Electric Canada, located in British Columbia, for 170 engines (three going to each helicopter), including spares, support, and training, was valued at over $108 million. Canadian Aircraft Products, also in British Columbia, won contracts as the prime international supplier of both the forward fuselage and the horizontal stabilizer. Initial orders were valued at over $12 million; however, the potential world market was valued in excess of $75 million, and countless years of employment were expected for many provincial residents.[31] IMP in Halifax was expected to be responsible for the engineering, mission system modifications, support equipment, flight tests, and integrated logistics support that would have provided approximately $400 million to the company. And EDO Canada in Winnipeg was selected as the prime international supplier of sponsons for the *EH-101,* with the potential world market valued at over $71 million.[32]

It would be naive not to acknowledge that the millions of dollars involved in the NSA contracts were in reference to the "potential" value of international sales. They were most likely inflated simply by the enthusiasm and marketing strategies of the companies involved. Notwithstanding their optimism, if the forecast assessments of these companies were significantly less than imagined, they would still have been a substantial addition to the Canadian gross domestic product. Perhaps equally important would have been the improved technological knowledge infused into the Canadian aerospace industry, which would have greatly enhanced the quality of Canadian skilled labour and capital resources. As NSA Program Manager Harvey Nielsen stated at the time, "these are high tech jobs – this is the stuff that can be exported out of the country. The fundamental characteristic of the economy is changing. Software and the ability to manufacture data – chips, high-density circuit boards – is what it's all about."[33] The IRB program was one of the primary reasons that the NSA program had taken so long to define and approve. The former program had made the latter program high risk due to the development of new technologies to be included in the final helicopter. But by 1993, everything was already contracted for and under way. The time to avoid the delays of inserting Canadian content had long passed. Following the 1993 cancellation, it was speculated that Canada's naval weaknesses would become a "burden for the rest of the allied fleet" and could preclude the navy from participating in international actions.[34] This opinion was predicated on the fact that an "increasing number of smaller navies are buying submarines because modern designs enable them to pose a significant threat, even to the most powerful navy."[35]

It was clear that EHI did not push the Canadian government as hard as it could have for punitive costs associated with cancellation of the NSA.[36] Ever since the release of the 1994 Defence White Paper, EHI had been aware that the DND was preparing to issue a Solicitation of Interest (SOI) to aircraft manufacturers in another attempt to replace Canada's SAR *Labrador* helicopters. In fact, the termination cost negotiations were completed just as the government announced that it was prepared to spend approximately

$600 million to buy fifteen SAR helicopters.[37] Canadian Minister of Defence David Collenette had already written to Malcolm Rifkind, UK Secretary of State for Defence, and Domenico Corcione, then Italian Minister of Defence, advising them that "there will be no impediment to the participation of Westland Helicopters Ltd. The Department of National Defence would welcome and gladly assess any innovative and financially attractive proposals."[38]

Although the 1994 Defence White Paper had made it clear that there was still a need for new maritime and SAR helicopters, the projects were once again split.[39] Doing so blurred the bottom line cost of replacing the cancelled NSA/NSH program as two projects were thought to look cheaper than one. In fact, this was part of the political direction of the 1994 SOR, along with the 25 percent capability reduction. The new SOR for the SAR helicopter was approved in June 1995, and on 8 November a DND news release announced that the department had finalized its SOR and that the Chrétien government intended to proceed with procurement of new SAR helicopters from a list of certified aircraft. The announcement also stated that the successful bid would recommend financing mechanisms and provide a maintenance component. Minister Collenette was quoted as saying,

> Buying off-the-shelf or alternative delivery arrangements such as leasing of equipment and contracting out maintenance brings with it the potential to reduce the cost of major capital acquisition projects. It will help us obtain value for taxpayers' money and still provide essential uninterrupted Search and Rescue service to Canadians through new and different forms of partnership with the private sector.[40]

The new SAR program, therefore, would follow the new policy of buying a certified aircraft off the shelf and avoiding the pitfalls of the NSA program. When the SOI, or what they called the "Statement of Interest," was released to industry on 29 January 1996, it referred to "the Government's intention to contract out all or part of the follow-on support, including maintenance, traditionally provided by DND."[41] Any company that responded, therefore, was asked to include what would soon become a common term in Canadian procurement – "life cycle cost data" – or what it would cost to maintain the aircraft over its estimated lifetime.

At the conclusion of the competitive process, the *EH-101* was again determined to be the best helicopter for Canadian requirements. It was a scaled-down model and was to be called the CH-149 *Cormorant*. In August 1997, after the DND and PWGSC realized that EHI would be the clear winner of the Canadian Search and Rescue Helicopter (CSH) competition, they informed

ADM (Mat) Pierre Lagueux. It was feared by the CSH Project Management Office that the government would challenge the results and try to avoid embarrassing the Prime Minister. Lagueux immediately directed the CSH project office to conduct a completely independent forensic audit of the CSH evaluation to confirm the results. KPMG Washington, the office responsible for all American Department of Defense procurement audits, quietly accepted a sole source contract from CSH Contracts Manager Ken Brown of PWGSC to conduct an Independent Validation and Verification of Bid Evaluation.[42] The redundancy contract was to be completed from 28 July to 14 August and would cost the government $132,290, "Goods and Services Tax Extra." KPMG confirmed the competition results, concluded that the CSH procurement had been conducted in a "fair, reasonable and consistent manner," and believed that "the CSH project was one of the best procurement evaluations we have seen in terms of documentation; electronic database; proposal evaluation procedures; and fairness, consistency and reasonableness."[43] The company added that the methodology of analysis for the CSH based on value was the best model it had seen and should serve as a model for future competitive defence equipment procurements.

After the KPMG audit, it was explained to Minister of National Defence Art Eggleton that the *EH-101* was still considered the winner of the competition. He then presented cabinet with the results. Cabinet then decided to spend more time and money to direct the Department of Justice (DOJ) to determine whether it could find a way out of choosing EHI. Following the initial DOJ report, which indicated that it would not be lawful to reverse the CSH competition results without good reason, the Prime Minister's Office directed that an independent legal opinion be sought. Retired Ontario Supreme Court Justice Charles Dubin was hired for this purpose. A letter from Ellen Stensholt, Senior General Council at Public Works, to Dubin revealed that he was to be paid up to $2,000 a day plus expenses from 11 December 1997 to 5 January 1998 to look at the project.[44] At the end of his analysis, Dubin concluded that there were potentially severe international and legal consequences if cabinet were to overturn the CSH competition results without cause. Finally, the Toronto law firm Lang Michener was also hired that December to see whether the competition could be discredited and restarted. The firm was offered the same rate as Dubin, and its conclusion was the same. The final bill for the law firm came to $16,382.37. Even a clean and well-run procurement, therefore, had become a waste of the Canadian taxpayers' time and money. But the NSA/NSH cancellation in 1993 had put the Liberal government in a tenuous position. It was a clear demonstration of irresponsibility to agree to pay nearly $500 million in termination costs, cut aerospace jobs, and reverse all its criticism of the *EH-101* by agreeing to buy it five years later. It is no surprise that the government tried to stop the procurement. But the effort to avoid admitting that the *EH-101/Cormorant* was the best helicopter for Canadian SAR failed. In January 1998, the Liberal government announced the award of a $593 million contract for fifteen *Cormorant* SAR helicopters to the company that it had so vociferously rejected in 1993. The Liberals disingenuously claimed that the *Cormorant* was a different helicopter than the *EH-101*. Most of the media fallout, however, was overshadowed by the severe ice storm that struck eastern Canada at the same time. But it was still significant. It made Chrétien acutely aware of how much more politically damaging it would be if they were ever required to award the *Sea King* replacement contract to EHI as well.

The SOR for the MHP continued in limbo throughout the remainder of the 1990s. On 20 January 1998, ADM Mat Pierre Lagueux sent an e-mail to the Vice Chief of the Defence Staff, Vice Admiral Garnett: "Now that a decision was made on CSH, questions are starting to be asked about MHP ... Given what we experienced with CSH, we better have our ducks in order if we hope to move this project successfully through approval process ... Timing and choice of procurement approach, as well as definition of operational requirements will be key."[45] This e-mail demonstrates the realization by the civilian official in charge of defence procurement that the maritime helicopter operational requirements and procurement strategy would need to be influenced from the top from the beginning. Defining any SOR for a major project takes considerable time as it eventually determines the project budget and the number of aircraft to purchase. But in the case of the MHP, it was the procurement strategy itself with which those involved were most concerned; they feared that the Liberal government would jettison the project again to avoid admitting that it had made a costly error in 1993 by cancelling the NSA. And they had learned from the CSH evaluation that, if they were to influence the outcome of the competition, such influence had to come before the Request for Proposal (RFP) was released to industry. A key component of shaping the procurement strategy was to continue to pare down the requirements to allow as many aircraft to compete as possible.

The week after the Liberal government announced the award of the CSH contract to EHI, the DND scheduled a meeting of MHP stakeholders, described as a "working group," to conduct a review of the 1995 maritime helicopter SOR to confirm that it was still valid. This meeting was held in Halifax on 12-13 January 1998. The group discussed how the original SOR had been based on the 1994 Defence White Paper and reiterated how it had included the statement "Canada needs armed forces that are to operate with the modern forces maintained by our allies and like-minded nations against a capable opponent – that is, able to fight alongside the best, against the best."[46] The working group confirmed that the maritime helicopter SOR was still valid, ensured that it stated the minimum essential requirements, and demonstrated that two or more aircraft types could meet the requirements as stated. The group also recommended changes to the SOR to "open and broaden" the competition. One of the documents presented to the group provided examples of the percentage reductions between the NSA and the MHP SOR for a number of operational requirements.

The endurance capability of the aircraft was of particular concern to some naval officers. In April, Captain E.J. Lerhe helped to write a briefing note drafted by the Canadian Forces Maritime Warfare Centre. He explained that his colleagues who had drafted the SOR should not be dropping the endurance capability lower than three hours and fifty-five minutes in the ASUW scenario. He also claimed that they "intentionally remained conservative" regarding littoral operations.[47] Despite these claims, the endurance requirement was continuously watered down from the mark of three hours and fifty-five minutes.

In June, Lieutenant Colonel David Neil, Project Director for the MHP, and Lieutenant Colonel R.F. Drummond, Project Manager, gave a briefing to Chief of the Air Staff David Kinsman and Chief of the Maritime Staff Admiral Greg Maddison on the continuing draft of the SOR.[48] Neil took them through a brief history of the technological obsolescence of the *Sea King* and the long and troubled attempt to replace it. The issue of new roles for the aircraft was also addressed, with importance being placed on the post-Cold War era and how the focus had gone from open-ocean ASW to peace support operations, the enforcement of United Nations resolutions, and humanitarian assistance. It was stressed that any replacement aircraft would need to operate not only in any domestic situation but also anywhere on the globe. Eleven extensive defence scenarios compiled from the 1998 *Defence Planning Guide,* from SAR to combat, were explained, as was how the new maritime helicopter would fit into each one according to a joint navy and air force working group. The example for SAR centred on one of the most recent missions, conducted in December 1995 800 nautical miles east of Bermuda, when the embarked *Sea King* of HMCS *Calgary* was called on to rescue thirty passengers from a sinking vessel. The helicopter was considered to have performed in a limited capacity regarding endurance and load and put the lives of the survivors at risk. As in the rest of the scenarios, the *Sea King* was considered deficient, and the new maritime helicopter would have to outperform its predecessor in all areas. The other ten scenarios included disaster relief in Canada, international disaster relief and humanitarian assistance, surveillance/control of Canadian territory and approaches, evacuation of Canadians overseas, aid to the civil power, national sovereignty/interest enforcement, peace support operations (UN Chapter 6 and 7), defence of Canadian/US territory, and

North Atlantic Treaty Organization (NATO) collective defence Neil made it clear that, to deliver these requirements on time, the new purchase policy would be to buy off the shelf and not use any developmental technology as had been done with the NSA/NSH program.

Clearly, the chosen helicopter would need to be of the highest quality to undertake all of these missions. The possibility of this was completely inconsistent, however, with the massive reductions to the requirements being set out by the draft SOR. Although it demanded that the helicopter be highly resourceful, companies that were incapable of providing such potential would still be asked to compete. And for all the working groups, defence objectives, scenario analyses, task assessments, and departmental approvals since 1995, the final SOR was still not completed. This was because, for the first time, senior military officials were insisting that direction for the procurement come from their political leaders instead of simply outlining their requirements for military operations. How the SOR was written was dependent on how the authors interpreted the uniqueness of their requirement, which affected whether it could be bought off the shelf. It was also crucial to decide whether the final cost was the dependent variable and whether IRBs were vital and at what level. All these factors would determine how many aircraft would be able to comply with the parameters for the contract competition.

After a briefing to the Deputy Minister of Defence and the Minister of Defence, Art Eggleton, on 7 December 1998, Pierre Lagueux sent VCDS Garnett an e-mail that day referring to the new process being considered: "These are not low level/PMO [Project Management Office] decisions. Direction must come from the top as you suggested this morning. Indeed, given the new process that is being considered, I would suggest that ministers (at least MND) have a direct say/concurrence ... that is quite different from any previous approach used in the past."[49] On 15 February 1999, a presentation on potential contenders for the maritime helicopter was given. The list of candidates was broad, and the advantages and disadvantages of each aircraft were outlined. The fourth slide of the presentation made clear that the AgustaWestland *Cormorant* had the possibility of "political fallout."[50] No other aircraft had this disadvantage. Although the Sikorsky *S-92* prototype again made an appearance, its disadvantage, as it had been during the NSA, was that it was still not certified in the naval version. The document needed was issued by an airworthiness government body, such as the Federal Aviation Administration (FAA), and certified that the aircraft met the applicable standards for the aeronautical product, proved that it conformed to its type design, and confirmed that it was safe to fly in Canadian airspace. The briefing concluded that DND needed a proven performance record and the option with the least risk.

After the first and second drafts of the revised SOR were rejected by the Senior Management Oversight Committee, a "Steering Group" was formed to ensure that the SOR described the minimum operational requirements.[51] At the end of the SOR review process in the summer of 1999, it was obvious that some in the Steering Group were disappointed with the final results. After an apology from Colonel G. Sharpe, Director General of Air Force Development, to Lieutenant General Kinsman for not being allowed to attend the last meeting with him, which had "less than satisfactory results," Kinsman replied,

> I'm not sure why the VCDS was off the mark (I don't think he had an agenda ... I think he just pooched it) ... I don't believe any notes you would have sent me or any comment you may have made ... would have changed the course of history (don't take that personally ... it's just that the Vice had his red suspenders on and he knew which set of assumptions he thought would give us the best chance to get a helicopter purchase underway.[52]

It was also that May that Deputy Prime Minister Herb Gray began to be briefed on the maritime helicopter procurement strategy and began to advise those responsible for the project.[53]

It was not until July 1999 that the downgraded SOR was finally approved. It was determined that four models from three separate manufacturers could meet the revised SOR – AgustaWestland, Eurocopter, and Sikorsky – and that twenty-eight aircraft would be necessary to fulfill the Defence White Paper commitments.[54] With the requirements finally approved, the maritime helicopter PMO turned to finalizing how the procurement would be carried out. As the SOR had been written with the actual procurement strategy in mind, the topics of IRBs, off the shelf, and a maintenance contract were very familiar. Part of the discussion in one of the early documents on acquisition strategies was whether to go against the Agreement on Internal Trade (AIT) – the only trade agreement that applied to the MHP – and sole-source the contract to one supplier. This could only be done under certain conditions, such as only one supplier being technically able; unforeseeable urgency; only one supplier by intellectual property ownership; compatibility of existing products on inventory; or procurement of a prototype developed for a particular contract. Invoking any of these clauses opened the contract up for challenge by the Canadian International Trade Tribunal (CITT). The other possibility was invoking the national security exemption, which would take the SOR out of the AIT and any recourse normally possible through the CITT. The only legal action possible, in that case, would be in a federal court. Any of these actions would mean no competition for the MHP contract, which was expected to cost between $2.5 and $3.5 billion. The downside to this was seen to be a lack of potential for IRBs. Another aspect of the possible procurement strategy was particularly important: whether the final contract would involve one contract or two – one for the helicopter airframe itself and one for the mission system inside. With the NSA, there was one prime contractor, and it was to find a company to inject a task-specific Canadian mission system. There was only one contract. The document concluded that this strategy reduced the risk to the government and that "experience suggests represents best overall value for money."[55] After all, the risks with having too many companies responsible for system integration were made clear with the NSA/NSH project. It took too long and cost too much.

As the process continued toward the SOI and the RFP, EHI took steps to communicate with the government regarding what it could offer. A letter written to the new ADM Mat, Alan Williams, explained the advantages that would exist regarding commonality if the government were to choose the *Cormorant* for the MHP role as well as the CSH role:

> There are a number of significant personnel, operations and maintenance savings which we believe need to be considered. There are:
>
> - A single nationally managed capital acquisitions program;
> - An abbreviated procurement process;
> - Common pilot and maintenance crew training;
> - Greater flexibility in the air and ground crew rotation among squadrons, and the associated reductions in sea time for helicopter squadron personnel;
> - Common flight simulators;
> - Reduced translation costs;
> - Reduced need for data rights and engineering drawings;
> - Common spare parts procurement and management requirements; and
> - Common third line repair and overhaul contractors.
>
> Apart from the significant initial procurement savings of approximately 5 percent (related for example, to fixed costs spread across a larger fleet, smaller project office requirements, shorter procurement process, spares procurement and management, training aids, publications, data rights, etc), our estimates suggest that the total life-cycle cost savings for a single fleet over a dual fleet would be in excess of C$500 million.[56]

The shorter procurement process was of particular importance as the *Sea King*'s airworthiness continued to deteriorate. On 2 December 1999, a *Sea King* experienced a single engine failure while attempting to land on the deck of HMCS *Protecteur* in Dili Harbour near East Timor and was forced to make an emergency landing in the water near the ship.[57] The pilots were fine, but the return of the aircraft to the ship had to await the conclusions of the investigation. The helicopters were also unable to carry out many of their tasks due to their inability to operate in hot climates. As a review of Canadian defence spending in 2000 by the Military Affairs and Defence Committee concluded, "each and every one is an accident waiting to happen."[58]

In 1994, the Canadian defence policy requirement of a multi-role, combat capable force was the same as during the NSA project, despite the Liberals' perception that the NSA was just for anti-Soviet submarine warfare. The stated requirements included protection of the fisheries, drug interdiction, environmental protection, humanitarian and disaster relief, and potential demands for aid to the civil power. The traditional role of ASW was also present. What had changed, however, was the will to ask for a robust helicopter to fulfill these roles due to the fear that it would be rejected once again over political parrying. That the NSA had been rejected by the Liberal government in 1993 made creation of the new SOR politically sensitive. Admiral Garnett was under orders to produce a document that would lead to the purchase of a less capable aircraft than the *EH-101*. Complete threat assessments had become extraneous to the process. Notwithstanding what the final SOR would look like, it was clear that NDHQ had decided to emphasize the purchase of equipment off the shelf. It was also determined to avoid implementation of domestic modifications. The IRB program was one of the primary reasons that the NSA program had taken so long to be defined and approved. The former program had made the latter program high risk.

In 1995, the total costs of termination were determined to be $478.3 million. In addition, the estimated $250 million investment in the defence industry was never given a chance to materialize. The day after the cancellation, a great many people lost their jobs. The termination costs were actually far less than had been expected, and the amicable negotiations were the result of EHI's desire to win the new competition for Canada's fifteen SAR helicopters that the company knew was about to be announced. The SAR and naval projects were again split to make the overall project appear cheaper to the public. And EHI did indeed win with a reduced version of the *EH-101*. But it was not easy. Selection of the *Cormorant* came as a surprise to the Liberals. Although the government went down every avenue imaginable to preclude the contract going to EHI, it was clear that it would have been illegal to do so. The selection of EHI once again created a commitment among the Liberals to avoid its being selected a third time for the MHP. Although the military had hurt itself regarding the creation of a diluted SOR, the government took it much further. And the CSH experience had taught the government how to do it. The key to influencing the outcome of any procurement was to intervene before the release of the RFP to industry.

A *Sea King* helicopter onboard HMCS *St John's* takes off for Chardonnière, Haiti, with her load of 1,000 kilograms of corn soy blend on 15 September 2008.

Source: CF photo by M.Cpl. Eduardo Mora Pineda. *Courtesy of National Defence. Reproduced with the permission of the Minister of Public Works and Government Services, 2009*

9
The Maritime Helicopter Project: Procuring on Eggshells

> As the prime example of why the procurement process must be reformed, the committee notes that a replacement for the Sea King helicopter – an operational need first identified almost 25 years ago – has yet to be delivered to the Canadian forces. It is the opinion of this committee that the failure to replace the Sea King helicopter epitomizes everything that is wrong with the procurement process.
>
> – Procurement study conducted by SCONDVA and presented to the House of Commons, 14 June 2000

On 17 August 2000, Art Eggleton and the Minister of Public Works and Government Services Canada (PWGSC), Alfonso Gagliano, announced that the government had given the Department of National Defence (DND) approval to proceed with acquiring a suitable replacement for the *Sea King* helicopter. Letters of interest (LOI) were subsequently sent out to possible contenders for the contract. The procurement process chosen by the government was to split the contract between a company that would build the airframe and another company that would build the mission systems and avionics inside. As a result, the project office then had to lay out two sets of requirements. The government refused to explain its rationale for choosing this strategy and went against the advice of the DND and PWGSC. It was later discovered that $400 million had been allocated to the budget to deal with the potential problems that this strategy created due to the increased risk of "contract omissions or errors."[1] But this decision effectively precluded the savings that were possible through AgustaWestland International Limited (AWIL) by the commonality of aircraft that Canada already used for Search and Rescue (SAR), specifically regarding life cycle costs. In fact, the split procurement allowed more companies to compete, which made it easier to avoid choosing AWIL, which had replaced EHI in July 2000. The *EH-101* had always been the strongest competitor in the attempts to replace the *Sea King*. The Liberals also realized that, if they cast aside the "best value" methodology used in the Canadian Search and Rescue Helicopter (CSH) project, it would be easier to avoid going with the *EH-101* based *Cormorant;* it was more capable, but it was also more expensive than its competitors. A "lowest-cost-compliant" matrix was then implemented, which essentially stated that, if one company's bid was a dollar less than the others, as long as it was deemed compliant, then it would be chosen regardless of overall quality and value. This approach contradicted the Treasury Board contracting policy: "Inherent in procuring best value is the consideration of all relevant costs over the useful life of the acquisition, not solely the initial or basic contractual cost."[2] It also forced the Maritime Helicopter Project (MHP) office to re-evaluate what would be considered a mandatory requirement in the Statement of Requirement (SOR). By 29 September, there were five interested bidders on the basic helicopter and thirteen interested bidders on the mission system and systems integration. All interested bidders confirmed that they could provide twenty-eight fully integrated maritime helicopters within the financial cap and the initial delivery date of December 2005. The final Request for Proposal (RFP) to all potential candidates was to be posted on the DND website by November 2001.[3]

AWIL soon protested the chosen procurement strategy of the Canadian government. G. Bologna wrote to the defence minister that

> It flies in the face of common sense not to recognize that operating one common fleet for both requirements is cheaper than operating two different ones. Surely the final definition of cost must include recognition of this fact or the government could find itself in the position of acquiring a cheap helicopter that adds significantly to DND's overall costs – nullifying any initial benefit.[4]

The company was convinced that it offered the solution with the lowest total cost for the entire life cycle and the one that would deliver the best value and come within the approved budget. But savings as a result of commonality of aircraft was never considered as part of the procurement evaluation. The split procurement was particularly troublesome to some analysts. Former *Sea King* pilot Colonel (retired) Lee Myrhaugen wrote presciently that

> in the proposed process, the "so-called" Basic Vehicle will be built and delivered according to specifications that cannot possibly take into consideration the size, weight, shape, power needs, etc., of a yet-to-be-determined mission suite. It is not made clear who will pay for the modifications to the airframe that the systems contractor requires in order to fit the mission suite equipment into the basic helicopter. Nor is it made clear who will do the

metal cutting and modifications and then re-certify the modified helicopter as airworthy once all of this is done. And who will pay for this? Moreover, who will be held accountable when all the extra modifications and work required result in unforecasted project and delivery delays?[5]

On 11 October 2000, EHI filed its first complaint with the Canadian International Trade Tribunal (CITT). The tribunal subsequently determined that EHI did not have a case. But when the company filed an appeal of the 31 October CITT determination, the Federal Court of Appeal indicated that it believed that the circumstances surrounding the MHP "may eventually demonstrate that the procurement procedure suffered from patent politicization within the Department of National Defence."[6] On 23 March 2001, Prime Minister Jean Chrétien also announced the hiring of David Miller as a special adviser. Miller was a registered lobbyist for one of the competing helicopter companies, Eurocopter.[7]

It was not only the CITT that was looking into how the MHP was being run. In April, the Auditor General, Sheila Fraser, wrote to Senator Michael Forestall in response to his letter dated 26 March regarding the MHP procurement strategy. Fraser wrote that her office placed an "emphasis on competition, full lifecycle costing and 'best value' contracting."[8] This policy was also made clear to Assistant Deputy Minister (Materiel) (ADM Mat) Alan Williams and ADM (PWGSC) Jane Billings by Fraser's assistant, David Rattray. He sent them both a letter the same day to express the essence of what the Office of the Auditor General needed both the DND and PWGSC to understand. His letter to Billings made it especially clear: "I still believe that it is important that I reiterate several of the basic principles that underlie our audits of government procurement. The most basic of these ... include competition, 'best value' procurement, and life cycle costing."[9] The attached slides made clear that it was against the contracting policy to focus on the lowest initial cost.

While those at the DND and PWGSC were being directed on procurement protocol, Raymond Chrétien, the Canadian Ambassador to France and the Prime Minister's nephew, sent an e-mail to the Prime Minister's Office, Minister Gagliano of Public Works, Deputy Minister Herb Gray, the Prime Minister's Chief of Staff Jean Pelltier, and senior policy adviser Eddie Goldenberg titled "Renewed French Misgivings." It explained that officials at the French company Eurocopter were having issues with the Canadian procurement process and were concerned that it was already behind schedule. One concern was that the aircraft had to be able to operate with one engine inoperable, which the company's two-engine model could not do. In fact, only the *Cormorant,* which had three engines, was able to comply with this requirement. Most of the e-mail released under the Access to Information Act (AIA) was severed due to its political sensitivity. But Chrétien did close his communication by expressing that "this is a tremendously important file from both the commercial and political perspectives."[10] Apparently, Raymond had his uncle's political sensitivity toward military contracts, and he used his position to attempt to recommend changes to the aircraft Requirement Specifications (RS) on behalf of a French company. The RS differ from the SOR in that they are the technical translation of the operational requirements outlined in the SOR; the RS document is used to measure a company's compliance with the RFP and award a contract. Making alterations to it was obviously outside the ambit of a political ambassador lobbying from France, especially one related to the Prime Minister.

Indeed, the RS were being diluted. This time it was by Colonel Wally Istchenko, Director of Air Requirements at the DND since July 1999. This position made him the custodian of the maritime helicopter requirements and the senior Canadian Forces (CF) person responsible for initiating or authorizing any changes to the RS. Some at the MHP office were not pleased with the reductions and warned that they could lead to a flawed procurement. The specification in question was the aircraft's ability to operate in international standard atmosphere (ISA) +20°C (flying in temperatures up to +35°C). Major N.B. Barrett from the MHP office, in an e-mail dated 11 April 2001, stated,

> I am writing to confirm that we will remove the "ISA+20C" performance conditions in lieu of an ISA standard day based on the direction given yesterday by Cols Istchenko and Drummond. For the record, I disagree with this direction ... Diluting to the lowest *un*common denominator ... will result in legal challenges from other contenders. Candidates meeting or exceeding the questioned requirement will see the specification dilution to the deficient contender's capability as bias towards the lowest performing competitor ... I am sure the greater operational community would take extreme umbrage at the procurement of a new vehicle where we ignore performance requirement in areas where they routinely will be operating.[11]

If the SOR was set at the "minimum operational requirements," then it made little sense to continue reducing them in the RS. As Barrett wrote, "to paraphrase, the logic is, 'you may add data [in the RS] where we forgot it unless we decide later that we didn't forget it.'" According to him, this action set a dangerous precedent and would surely bring logical protests of bias toward the lowest-performing competitor from the other companies meeting or exceeding requirements such as the ISA+20°C operability. It was setting a requirement (flying at ISA+15°C) that was less than what the *Sea King* could perform. The other possible option was that the more capable contenders could reduce their performance specifications, adjust their cost baselines, and offer the additional performance capability at a cost after their particular vehicle had been selected. Colonel Henneberry, Commander of the Maritime

Air Component (Pacific), concurred with Major Barrett in a memo stating that the consequences of accepting the reduced ISA criteria "could severely impact our operational capability and is clearly not in the best interest of meeting the needs of the Canadian Forces."[12]

By the summer of 2001, the Liberal government was being challenged on its political direction of the MHP. That the procurement strategy itself did not come from the procurement experts at the DND was made clear in testimony by Alan Williams to the Standing Committee on National Defence and Veterans Affairs (SCONDVA) on 5 June. Williams was the senior official responsible for the bidding process.

Leon Benoit:	Whose decision was it not to bundle [the procurement]?
Alan Williams:	The decision came to us from government.
Leon Benoit:	So it was the government's decision to do that, and you will admit it does increase risk?
Alan Williams:	Yes.[13]

The government, therefore, was cleverly sidestepping what some military professionals were recommending as to how the project should be undertaken. And Chrétien attempted to maintain the illusion that his office was not providing direct instruction on how to carry out a procurement competition. In response to a question from Joe Clark during Question Period on 6 June, Chrétien denied the existence of the Gray committee: "Here is the most honourable member in this House of Commons – and he is not chairing any committee on this program, and if he were, I know that he will do it with competence and honesty."[14] The very next day Chrétien was forced to admit the truth when pushed by Clark: "As Deputy Prime Minister he presided over a committee to look at the process of establishing the bids that are out at the moment for people to make submissions." Although Gray denied the existence of the committee the following week in Question Period, the same day he also stated, "yes, we looked in the past at how this process might operate. The work that we did in the past resulted in an open and transparent process."

During conversations with the possible candidates regarding the RS, complaints were made by Eurocopter and Sikorsky, and it became clear to the MHP office that only the *Cormorant* could meet the RS without significant design changes. Colonel Istchenko addressed the situation in an e-mail: the MHP office could reduce the RS but leave the "essential items in the SOR intact and give some room to those potential bidders that are facing weight [carrying capacity] challenges."[15] The office prepared a list of proposed reductions to the RS for approval by 12 Wing. The DND calculated that it could drop overall carrying capacity by allowing the new helicopters to carry, among other things, smaller sonobuoys (45 kg less); one less life raft (25 kg); one torpedo instead of two in hot weather (225 kg); and no spotlight (30 kg), speaker (27 kg), or cargo hook (18 kg). Senior command at 12 Wing then undertook to review the possibility of reductions and return their recommendations to Colonel Istchenko, who had become Deputy Program Manager in the MHP office in December 2001.

In May 2002, Eurocopter withdrew its intention to bid on the MHP with the *NH-90*. Olivier Francou, NH Industries spokesman, said that the model was withdrawn because it did not meet the contract requirement for Canadian investment.[16] The DND required bidders to lay out a plan to invest the amount of the contract in industries across Canada. NH Industries built its airframes and onboard mission systems and weapons suites in Europe. For this competition, it would have had to ship $400 million of its mission systems manufacturing to Canada, said Francou. "We worked over the last two months with Canadian companies ... but it was almost impossible."[17] NH Industries then chose to work with Lockheed Martin and its Canadian subsidiary to make a joint bid on the MHP.

The government had initially stated that its objective for delivery of the first fully integrated, mission ready, maritime helicopter was no later than December 2005. The split procurement and the lowest-cost-compliant matrix had pushed delivery back to 2008. But on 5 December 2002, only two weeks after Alan Williams had written to AWIL to express that the government remained committed to the split procurement process, John McCallum, who had become Minister of National Defence in May of that year, announced that his government was reversing the procurement strategy and stated that there would be only one RFP for one contract for the airframe and mission system: "This action, which is consistent with the industry consensus, is a major step forward in getting lower risk for the taxpayer, better value for the Canadian Forces and ... speedier development of the helicopter."[18] Although the intention was to return to a more efficient process and reduce risk to the project, it meant another round of pre-qualification discussions with potential companies and an unknown date for the final RFP. It also raised many questions about why the government had chosen that strategy in the first place. Deputy Prime Minister John Manley said it best in his response to a question on political interference in the MHP during the Liberal leadership race: "There are lots of ways procurements get jiggled around."[19] Indeed, the leadership race allowed handling of the *Sea King* replacement project to take centre stage and made the candidates address the issue much as it had done during the Conservative leadership race a decade earlier. Paul Martin also made a statement regarding the MHP:

> Look, there is no doubt that we need the new helicopters now, and as far as I am concerned we should place the order as quickly as we possibly can.

> Also the question then comes down to the implication of the manipulation – the nature of the replacement. It should be absolutely the best helicopter that we can get based upon what the military themselves decide they need with the greatest capacity and the greatest variability, because one doesn't understand we can't project today what demands are going to be placed on them.[20]

With alteration of the procurement strategy, the media again awoke to the topic of replacing maritime helicopters in Canada and began to criticize the process. Senior defence analysts made it clear that they believed changes had been made to the MHP RS to circumvent the possibility of choosing the *Cormorant*.[21] Even Sikorsky, thought to benefit from the reductions, complained that they catered to NH Industries. Lloyd Noseworthy, Sikorsky's regional director of business development in Canada, stated, "I'd have to say right now that there's a lot of rumour out there right now that they've dropped the bar significantly, and we've certainly questioned the latest release of the specifications."[22]

Indeed, there was reason for criticism. The most damning of all analysis became public in May 2003. Colonel Brian Akitt, a former *Sea King* pilot and Commander of 12 Wing, had written a paper on the MHP while attending the National Security Studies Course at the Canadian Forces College. At the time, there were few military officers who had a more extensive understanding of the history of the attempts to replace the *Sea King*. In 1982, Akitt was posted to Director General Aerospace Engineering and Maintenance in National Defence Headquarters (NDHQ) as the initial project officer on the *Sea King* Replacement Project. By 1985, he was posted to the Directorate of Maritime Aviation at NDHQ as the staff officer for the *Sea King*. After resuming operational and command duties in the field, he returned to Ottawa in July 1996 to the Directorate of Air Requirements as the section head responsible for maritime helicopter requirements and as the project director of the MHP. His paper was leaked to the media and became a high-profile topic in Ottawa. The findings of the study were obvious from the title: "The Sea King Replacement Project: A Lesson in Failed Civil-Military Relations."[23] Akitt described the project as an "abject failure."

Within the study, he astutely placed blame on the military for failure to replace the *Sea King* and explained that the CF had failed to convince the government of the requirement and any urgency to replace the helicopter. The true problem, however, according to Akitt, was with the relationship between the politician and the military expert, most famously described by Samuel Huntington in his 1957 classical work on civil-military relations, *The Soldier and the State*. Akitt concurred with Huntington that ministers have an obligation to listen to the military professionals who define threats to the nation and the appropriate responses.[24] Notwithstanding the degrees of advice given or accepted, in Canada the civilian authority always reserves the right to make the final decision.[25] And in this case, as Akitt correctly asserted, Canadian ministers ignored their military advisers to ensure that the procurement would have a politically acceptable outcome: "With the MHP announcement of the procurement strategy as outlined on 17 August 2000, it is clear that the civil component of the civil-military relation had over ridden the military component, effectively negating any influence that the military could hope for in the outcome of the project."[26] Akitt concluded that, since the government did not trust those at the DND to state requirements that would preclude the *EH-101* from winning for the third time, it took control of the procurement strategy to do so itself.

In August 2003, the Advisory Committee on Administrative Efficiency reported to the Minister of National Defence: "Procurement is universally viewed as being a slow and cumbersome process that does not fully respond to Defence's (Department of National Defence and the Canadian Forces) needs."[27] They pointed out that the average time period for a major capital equipment program was much too long – fifteen years – and that the DND did not have an effective approach to risk management. McCallum maintained, however, that the Maritime Helicopter Project was his department's top investment priority and that the helicopters needed to be procured as quickly as possible. He explained that the onus for a "speedy delivery" lay with his department and that of PWGSC. He assured his audience that he would appeal to "the self-interest of those companies with a plan that awards bonuses for early delivery and imposes penalties for late delivery."[28] In fact, early delivery was prohibited. The government had already stipulated that it would not take delivery of a new helicopter until forty-eight months after contract award (MACA). Lieutenant Commander Dave Scanlon, a DND spokesman, explained that this stipulation was based on consultations with the aerospace industry and that DND needed time to train pilots and build new facilities to handle the helicopters. AWIL had already assured the government that it could deliver the first helicopter to Canada within thirty-five months and have all aircraft on the runway by 2008-9. Lieutenant General (retired) Larry Ashley, a former head of the air force and a retired *Sea King* pilot, called the four-year timeline "ludicrous": "Usually (a customer must) pay a premium to get something accelerated. If a guy says to me, I can deliver it in six weeks and the other guy says it will take six years, give me the six weeks any time."[29]

With deliveries scheduled to start in 2008, the air force had to figure out how to keep the *Sea Kings* operational. After all, they had just turned forty years old. Art Eggleton, the Defence Minister in 1999, had admitted at the time that the military would have little choice but to mothball the helicopters by 2005.[30] The Auditor General had noted that the availability of the *Sea Kings* for service in 2000 declined from 42 percent to 29 percent.[31]

Indeed, they had also suffered another series of serious accidents and malfunctions. On 23 June 2000, a *Sea King* crashed into the sea off the coast of Hawaii due to an overheated main transmission gearbox.[32] Eggleton, in Hawaii at the time to observe naval exercises, was subsequently transported around by an American helicopter.[33] The following month two *Sea Kings* were called on to intercept and board the GTS *Katie,* a cargo ship that had refused to return $223 million worth of Canadian Forces equipment being transported back from Kosovo because of a contract dispute with the government. One *Sea King* became inoperable, and the crew had to rely on the personal cellular telephone of one of its technicians to call the base for help. The lone functioning chopper had to make two trips to lower the fourteen-member boarding party.[34] On 4 September 2002, *Sea King* pilots were forced to make an emergency landing south of Halifax harbour and later discovered metal shavings in a gearbox.[35] In February 2003, a *Sea King* mysteriously lost power and crashed into the flight deck of Her Majesty's Canadian Ship (HMCS) *Iroquois* after attempting to take flight.[36] Two people were injured, and the destroyer had to return to port because of damage to the ship. The ship had just recently left Halifax and was on its way to the Persian Gulf to become the command and control vessel of eight warships involved in Operation Apollo. *Iroquois* could not fulfill its responsibilities on time because of the accident and was eventually sent back to the gulf without a helicopter. The government then had to pay $180,000 to send an alternative aircraft to the ship by a commercial vessel. There was one *Sea King* operating in the area aboard HMCS *Montreal,* but it needed to return to Halifax for maintenance. Part of the delays in keeping the *Sea King* fleet in the air also centred on the fact that the air force did not have enough trained helicopter mechanics to keep pace with the amount of work needed. The long hours needed to work on the helicopters had created a high rate of attrition and burnout that could not be replaced.[37]

A few days after the crash aboard *Iroquois,* the formal notice from companies that intended to bid on the MHP was extended by three weeks to 21 March to ensure that there would be at least three potential suppliers for the project. From May to October 2003, the potential MHP bidders – AWIL, Sikorsky, and Lockheed Martin – were required to submit a fully compliant pre-qualification proposal in order to receive the final maritime helicopter RFP. This stage required that bidders provide proof of compliance for specified requirements that amounted to approximately 15 percent of the RS. During the iterative process, it became evident to both AWIL and Lockheed Martin that the proof being asked for to demonstrate compliance, such as detailed test reports and videos for things such as blade and tail fold that the *S-92* did not possess, would be impossible for Sikorsky to provide.[38] It was speculated that Sikorsky could not possibly be found compliant. The original pre-qualification process as stated by the DND had been specifically constructed to eliminate bidders that could not produce their products on schedule.[39] After all, the idea from the beginning of the procurement was to avoid a developmental aircraft and purchase off the shelf. It was supposed to be already in use by another navy. This requirement was amended after it became clear that Sikorsky did not have an existing helicopter. As part of the pre-qualification phase, the process then permitted bidders to promise that, to the extent that their aircraft did not exist, it would be developed for Canada.[40] Sharon Hobson reported that: "The Maritime Helicopter Project started out with 1,400 mandatory technical requirements but in order to speed things up during the bidding process, the project office only required that the bidders provide proof of compliance for 475. The bidders were allowed merely to state that they would comply with the other 1,000."[41]

On 16 December 2003, PWGSC notified the bidders that only Sikorsky's and AWIL's pre-qualification proposals had been evaluated as being fully compliant. The announcement that Sikorsky had been evaluated as being compliant while Lockheed Martin was not came as a surprise to both Lockheed Martin and AWIL.[42] The RFP was released to the two remaining companies the next day. Paul Martin had just been sworn in as Prime Minister the week before. The government of Jean Chrétien, therefore, had successfully avoided being involved in the possible selection of the same helicopter that it had rejected in 1993. As Doug Bland had correctly predicted in 2002, "it's all about avoiding embarrassing the prime minister ... When the PM retires, the military will get new helicopters. It's as simple as that."[43]

On 14 May 2004, the RFP was closed, and evaluations for the selection of the *Sea King* replacement began three days later. By the summer, AWIL attempted to explain to the government that it was not possible for Sikorsky to deliver its aircraft within the mandatory forty-eight months after a contract was signed. AWIL wrote directly to the Ministers of PWGSC and National Defence and even took pages out in the *Hill Times* to explain to parliamentarians the risk associated with choosing Sikorsky. One article entitled "Will the Sea Kings Still Be in Service in 2011 and Beyond?" explained that the *H-92* – what the company was by that time calling the incomplete military version of the *S-92* – was to use the same Data Management System (DMS) as the one proposed for the upgrade of Canada's *Aurora* patrol aircraft fleet.[44] Sikorsky had already admitted this in its advertisements in the *Hill Times*.[45] The DND was already aware that development of this DMS was years behind schedule. This information, combined with the fact that the airframe itself was still not certified, should have made Sikorsky's bid non-compliant. As one member of Team *Cormorant* explained to me, "it's not that we were that much smarter than everyone else – it was just that obvious."[46] But Sikorsky gave its guarantee of delivery in 2008 nonetheless, and the government and DND accepted it.

On 23 July 2004, Minister of Defence Bill Graham and Minister of Public Works Scott Brison announced that Sikorsky's H-92 *Cyclone* had been selected for the MHP. On 23 November, Sikorsky was awarded two separate but interrelated contracts. The first, with a value of $1.8 billion, covered acquisition of twenty-eight fully integrated, certified, and qualified helicopters with their mission systems installed. The second contract, worth $3.2 billion, was for the twenty-year In-Service Support (ISS) for the helicopters and included construction of a training facility as well as a simulation and training suite.[47] L3 MAS, based in Mirabel, Quebec, was chosen to work on the long-term In-Service Support (ISS) of the aircraft.[48] According to the DND, the MHP evaluation was "based on proposals meeting mandatory requirements and achieving minimum passing marks on the rated requirements established in the evaluation plan. The selection process identified the bidder who submitted a technically compliant bid, at the lowest price with acceptable delivery, terms and conditions and Industrial and Regional Benefits."[49]

General Dynamics Canada (GDC), based in Ottawa, was chosen to be the mission system integrator. As with the New Shipborne Aircraft (NSA), a prime contractor was chosen to build the airframe and a Canadian subcontractor was to design, develop, and integrate the mission system. The contract included responsibility for the radar, acoustics, electronic countermeasures, DMS, navigation, communication, and weaponry. GDC had already acquired some experience through a separate subcontract for the Aurora Incremental Modernization Program in 2002. The company was to develop the mission computer systems for the *Aurora* and a similar system was to be integrated into the *Cyclone*. It was also responsible for the ISS of that system. The contract operated "on a performance-based availability contract framework."[50] Sikorsky was required to deliver the first fully capable helicopter no later than 30 November 2008, as per the forty-eight MACA requirement. The remaining helicopters were to be delivered at a rate of one per month thereafter.

Alan Williams praised the selected company at a meeting with the media: "Special thanks to Sikorsky who have proven through their comprehensive and professional bid that they can provide the platform, mission systems, and support apparatus the Canadian Forces require at the best price ... Sikorsky's submission demonstrated their commitment to the government of Canada and you can be certain that we will hold them to account."[51] In response to a question about whether it was wise to choose a civilian helicopter that was still to be converted, Williams noted that "any helicopter that we would have acquired would have to be tailored to meet our needs. The frame is one that is in production. The systems that will be integrated by [General Dynamic Canada (GDC)] are all commercial off-the-shelf products ... so we have absolutely no doubt that it can be done." The idea that having commonality of aircraft in the Canadian SAR and naval roles with the *Cormorant* would have saved money was dismissed because, as it was explained by the Assistant Chief of the Air Staff, Major General Richard Bastien, they are two very different roles. Even advantages of common training and servicing were dismissed.

It was also revealed that Sikorsky had promised $4.5 billion in Industrial and Regional Benefits (IRBs) if chosen. The DND explained that

> The direct benefit of this activity for Canadians will continue long after the delivery of the last helicopter, with work on the helicopter project continuing over the next 20 years. Sikorsky has committed to partner with 170 firms, both large and small, and from our Aboriginal business community, with most regions of the country being home to significant portions of the project activity ... Equally significant is Sikorsky's commitment to involve Canadian small business in work on the helicopters. Indeed, its winning bid includes fully $685 million in industrial activity to be undertaken in our small businesses.[52]

Williams was also clear that the SOR document had always been the basis of the procurement and that it – in its final form – was not tampered with to allow for certain companies to perform better.[53] But the downgrading of requirements actually began with those at the DND who knew that they had to deliver a SOR that focused on a minimum of operational requirements. The ghost of the *EH-101* "Cadillac" prevented them from asking for what they actually needed. The caution exercised in creating this document is clear from its constant revision. After all, work on it had begun in 1995 and was not completed until 1999. After that, further alterations were made to the Requirement Specifications (RS), not the SOR, and allowed Williams to claim that the SOR created by the military had never been changed. Indeed, some of the alterations to the RS were completely logical, as industry explained during the pre-qualification phase, since certain new technologies did exist and were more capable.

When the question was inevitably posed regarding the political embarrassment and waste of tax dollars that would have occurred had the government chosen the same helicopter that it had rejected in 1993, Williams said that they had nothing to do with it. If they had, then why would the government have chosen EHI for the SAR contract in 1998? In fact, it did not have much choice in the matter and had done everything in its power to avoid the controversy of choosing that helicopter. Only after it was advised by several third-party analyses that it had no legal alternative but to respect the competitive process did the government proceed with the purchase. Although the Liberal government was embarrassed by the deal, the *EH-101* variant was the best aircraft, and politicians could not reverse its selection at that point. But most people were unaware of these events. And the government had learned its lesson. It became aware that it would have to get involved in the procurement of the naval version far earlier – especially before the release of the RFP.

After the announcement was made that Sikorsky had won the competition, it was revealed to AWIL that its own bid had been determined to be non-compliant. This was never mentioned at the NDHQ briefing. There were four proposal parts that bidders had been required to submit: (1) a general contract proposal; (2) a financial proposal; (3) a technical proposal; and (4) an IRB proposal. The technical proposal had to be made up of twenty-two specified technical compliance plans. Each plan was scored for its response to a number of questions relating to compliance with the many technical requirements contained in the RFP contract documents. All six groups had to receive an aggregate score of 60 percent each for compliance, capability, and risk; otherwise, the bid would be disqualified. AWIL was assessed a score of less than 60 percent on two of the six technical proposal groups and was disqualified.[54] The specifics of this disqualification remained confidential. Gloria Galloway reported in the *Globe and Mail* that the *Cyclone* was a less capable aircraft, but Sikorsky's bid was approximately 1 percent lower than that submitted by AWIL.[55] She quoted her source as stating that "the difference in price was razor thin but the Cormorant bid was found unacceptable." The source had refused to divulge the bottom line on the *Cormorant* contract. In fact, the exact bid of AWIL was, and still is, confidential information. The Canadian taxpayer has never truly discovered how much was saved by choosing Sikorsky. The final determination, however, was not based simply on lowest cost. But Sikorsky understood well the terms of the bid evaluation. It knew that it also had to meet the other requirements, which included delivery of the aircraft on time. So Sikorsky promised to deliver twenty-eight *H-92s* – which existed only on paper and without a mission system or certification – in the same time period that AWIL promised to deliver its *Cormorants,* which required very minor modifications to the already operational and fully certified *EH-101*. The MHP evaluation did not assess a bidder's actual ability to deliver the specified goods but required only that the bidder make a promise, or "paper certification," that it could do so. One of the primary questions that arose at that time – since everyone involved in the MHP knew that the AWIL bid would be higher – was why bid at all? The answer was that AWIL was convinced that Sikorsky could not deliver on time and that, therefore, AWIL had the only compliant bid.

As part of the contract, the government also decided that, if Sikorsky could not deliver its aircraft on time, there would be a cap on how much the government could charge for damages. Public Works officials explained that there would be a penalty of $100,000 for every day delivery was delayed. That penalty was also capped at a year. This amount had actually been reduced from the original $250,000 a day set out in the RFP in December 2003.[56] Sikorsky, therefore, would not have to pay more than $36.5 million in late fees regardless of how long the helicopters were delayed. This was a relatively minor sum in relation to the size of the contract.[57]

Criticism from the media over selection of the *Cyclone* was immediate. David Pugliese, the most prominent Canadian journalist on Canadian procurement, quoted a defence review document from 1997 noting that the "selection of common airframe meets a DND objective to rationalize and reduce aircraft types within the CF thus reducing the operations and support bill ... There is also merit to being part of a large worldwide fleet of helicopters that may offer some support savings through co-operative ventures with user-groups."[58] Pugliese reiterated what AWIL had said all along: the *Cormorant* could have saved upward of $500 million on the support contract alone. The most troubling aspect remained that the *H-92* was not a finished product. As I wrote the week after the announcement,

> Governments truly hate to admit they were wrong. In order to avoid this, the new Canadian government has just bought an aircraft that does not exist ... While this seems to be the final conclusion of the most absurd procurement project in Canadian history, it is more likely that it is merely another frustrating chapter ... There is a powerful suspicion that the Cyclone will not be delivered by 2008 as stated in the contract.[59]

The new government, however, maintained that the procurement had been conducted fairly. Minister of Defence Bill Graham declared that "the suggestion that this [the selection of the *Cyclone*] is politically motivated – I don't know where this is coming from." Sikorsky also helped to stave off criticism as Lloyd Noseworthy, the company's regional director for international business, stated that revamping the *S-92* into the military variant would be relatively easy and that he had no concern about delivering it on time: "We have built probably more aircraft than all the other helicopter manufacturers put together, so this is business as usual for us."[60]

By August 2004, AWIL had reviewed the limited number of files released to it on how the MHP office had evaluated its bid. AWIL subsequently maintained that, in virtually all instances, the information said by the evaluators to have been missing from its proposal was in fact provided, either in the capability reports or in the proposal plans. The company also identified a series of factual errors in the evaluation of the bid.

On 1 September, AWIL brought allegations of political interference in the MHP to the Federal Court of Appeal in Ottawa. AWIL alleged "serious errors" and political favouritism in awarding the MHP contract.[61] It applied for "a declaration that the Minister [of PWGSC] has exceeded his lawful authority by designing and conducting the Procurement to serve an irrelevant and improper purpose, namely to avoid the political embarrassment that would have resulted to the Minister and his partisan interests if AWIL were successful in the Procurement." The eighteen-page document also revealed that the third party that the government employed to ensure fairness in the evaluation of bids was a registered and paid lobbyist that targeted PWGSC and the DND during the term of the procurement on behalf of GDC, one of the principal partners in the Sikorsky bid. He was also a former long-time associate of the officials at PWGSC and the DND who supervised the evaluation of bids. The document also maintained that the Sikorsky bid was truly the one that was non-compliant and that the Minister of PWGSC was aware of that fact. In particular, it asserted that he was aware that Sikorsky could not deliver the helicopter by the mandatory deadline stipulated and that Sikorsky had deliberately attempted to deceive the government. AWIL made it clear that the civilian *S-92* faced major design problems to operate on a ship before it would even be ready to carry a military mission system and evolve into the *H-92*. The lack of any evaluation of risk for deadline compliance and the institution of nominal penalties for late delivery allowed this possibility.

On 15 January 2004, a formal policy for the use of a fairness monitor on public procurements had been adopted by PWGSC. The policy stated that "a critical element of this [policy] will be to ensure that the fairness monitor is and is seen to be independent of those responsible for the specific acquisition." The Minister of PWGSC, however, chose André Dumas and his firm AML Associates Incorporated. As AWIL pointed out, the conflict of interest that violated the fairness policy was that AML was a registered and paid lobbyist of GDC as recently as March 2001, seven months into the formal procurement, and had lobbied PWGSC and the DND regarding the Combat System Engineering Support project. GDC, of course, was a partner with Sikorsky in the MHP bid. Dumas was also a former PWGSC employee who was a long-term associate of those conducting the procurement. AWIL, therefore, contended that there was a reasonable apprehension of bias in the evaluation of the procurement: "The evaluation included decisions by the Minister's officials to ignore instances of non-compliance in the bid submitted by Sikorsky and to purport to identify deficiencies in the bid submitted by AWIL."

Although lawyers representing the DND and Sikorsky Aircraft Corporation subsequently argued that the issue should not be heard in court, it took a panel of three Federal Court of Appeal judges less than ten minutes to decide that the case could proceed.[62] AgustaWestland later added a $1 billion claim for damages.

Composite image of the new CH-148 *Cyclone* helicopter. The *Cyclone* will replace the CH-124 *Sea King* as Canada's main ship-borne maritime helicopter.

Source: Photo copyright Sikorsky Aircraft Corporation. *Courtesy of National Defence. Reproduced with the permission of the Minister of Public Works and Government Services, 2009*

10
The *Cyclone* Decision: *Caveat Emptor*

> What is most important, however, is that acquiring an existing, proven technology greatly reduces the risks associated with acquisition. Operational performance can be evaluated and proven, there is more cost certainty in acquisition price, delivery dates can be defined with precision, and in-service support costs are more predictable.
>
> – Assistant Deputy Minister (Materiel) Dan Ross, 2007

On 23 November 2004, the Maritime Helicopter Project (MHP) contract was signed between the Canadian government and Sikorsky. Only a few months after the award, the government initiated a contract change, at considerable additional cost to Canada, for Sikorsky to design, develop, and deliver a tactical transport capability for the *H-92*. Of the final two MHP competitors, only the *Cormorant* possessed a designed, certified, and in-service tactical transport capability; however, this capability was neither considered nor given value during the competition because it had not been required by the Request for Proposal (RFP).[1] This addition created the necessity to modify project schedules and contract delivery times. By that summer, the MHP office was still firm that Sikorsky would deliver by 2008. In an interview with Sharon Hobson of Jane's Navy International, Colonel Wally Istchenko, Deputy Project Manager of the MHP, reiterated that all of the systems were "virtually off-the-shelf" and that they just had to be integrated.[2] In the next paragraph of the article, however, Hobson revealed that Sikorsky was developing a tail and rotor fold system for the aircraft and that it was still not certified to fly by wire, which would replace manual control of the aircraft with an electronic interface.

In early 2006, AgustaWestland International Limited (AWIL) brought forth more allegations of bias regarding the MHP. The company had just discovered that Colonel Istchenko had taken employment with General Dynamics Canada (GDC) a month before he officially retired from the Department of National Defence (DND) in December 2005. The DND post-employment policy prohibited senior members of the DND from obtaining employment with companies with which they personally, or through their subordinates, had significant official dealings during the year prior to the termination of their employment or service. As AWIL discovered, Istchenko had applied for an exemption from this rule, and it had been granted by Colonel J.L. Milot, Director of the Defence Ethics Program, because Istchenko had claimed to have had no official dealings with GDC. In April, AWIL filed an Amended Statement of Claim pertaining to Istchenko's employment at GDC that stated "the Minister of DND granted a special exemption to the senior officer of the Canadian Forces [CF] involved in the acquisition of the MH [maritime helicopter] to take immediate employment with GDC in contravention of the DND Post-Employment Code."[3] AWIL claimed that Istchenko and his subordinates had direct contact with GDC regarding contract deliverables for the MHP Integrated Mission System. The crux of AWIL's claim was that, throughout 2005, in his official position as a member of the Configuration Control Board within the MHP, Istchenko would have been involved in approving late delivery dates for preliminary design reviews, many related to GDC and their mission system.

Sikorsky was also having troubles with its workers. In February 2006, employees walked off the job because the company wanted them to pay 20 percent of their health care costs and double their current monthly insurance premiums.[4] The Teamsters strike ended on 6 April. As a result, Sikorsky was later granted a six-week excusable delay on delivery of the first *Cyclone* to Canada.[5]

Although the court proceedings against the government were set to continue into January 2008, it was announced that AWIL settled its dispute with the government out of court on 26 November 2007. This was done without any payment to AWIL. The lawsuit was dropped with little fanfare in what David Pugliese described as "the most overlooked defence story of 2007."[6] Alan Williams asserted that AWIL never actually had a case and finally realized it. This is highly unlikely. The company had already spent three years of effort, at considerable expense, and would not have done so if it was not sure that it had a chance to win. Indeed, as the official release by PWGSC stated, "AgustaWestland stands behind the allegations in the lawsuit."[7] A far more plausible reason for the cessation of legal proceedings was that AWIL wanted to remove possible obstacles to future business with the Canadian government. This reason was confirmed by Gordon Cameron, the lead counsel for AWIL: "AgustaWestland has identified two recent proposals by Canada to purchase military equipment ... AgustaWestland is interested in competing in those procurements without any impediments such as might be posed by the litigation."[8] A potential contract for AWIL could involve armed escort helicopters for the air force's fleet of *Chinook* transport helicopters being procured at the time. Even the PWGSC release acknowledged the possibility of future contracts with AWIL: "The Minister of Public Works and Government Services and the Minister of National Defence welcome AgustaWestland's participation in any future military procurement."[9] In fact, this was the second time that the company had let the Canadian government off easy in the hope of continuing their business relationship. It had agreed to a cost of termination much lower than expected for the New Shipborne Aircraft (NSA)/New Search and Rescue Helicopter (NSH) project in 1996 in order to put itself in a better position to win the competition for Search and Rescue (SAR) helicopters announced the same year.[10] Whether it was a factor or not is unclear, but they did win the contract.

Although AWIL's interest in future business with the Canadian government was an important factor in dropping the case, there was more to the story. A Supreme Court decision in the case of *Double N Earthmovers Ltd. v. Edmonton (City)* in 2007 had a direct effect on AWIL's decision. The original dispute arose from a call for tenders by the City of Edmonton in 1986 for waste removal services. The city considered two seemingly compliant bids from Double N and Sureway. A key stipulation in the competition was that the equipment to be used could not have been built before 1980. Although Sureway offered the lowest bid, Double N pointed out to the city that some of Sureway's equipment was pre-1980. The company demanded that the city investigate and consider Sureway non-compliant. The city did nothing and gave the contract to Sureway. Ten days after the contract was signed, Sureway revealed that its bulldozers had been manufactured in 1977 and 1979. Although the city demanded that the company use compliant equipment after the contract was signed, Sureway explained that it could not do so. The city accepted the situation to prevent further confrontation. Double N sued the city, and the case entered the legal system for the next twenty years. The final decision by the Supreme Court was five to four in favour of the City of Edmonton. The court concluded that

- The owner's obligation to treat bidders fairly and equally ends once the winning bid is accepted; and
- Owners are entitled to assume that a bidder will comply with the terms of its bid. Owners can limit their evaluation of a tender to its face, so that there is no duty to investigate whether a bid is compliant.[11]

According to Paul Lalonde, co-chair for the International Trade and Competition Law Group at Heenan Blaikie law firm, this decision violated the fundamental principles of twenty-five years of Supreme Court jurisprudence in tendering law. In all the previous rulings, the owner had an obligation to accept only compliant bids. In short, the new ruling interpreted the tendering process in a way that rewarded a bidder who deliberately made false statements. As Lalonde wrote, "[contracting] analysis has been severely undermined and collusion between owners and preferred bidders has been made easier." According to this legal precedent, therefore, even if Sikorsky intentionally submitted an unrealistic bid in the MHP competition, once the contract had been signed there was no recourse available to AWIL. As Colonel Laurence McWha has explained, "if the customer knowingly accepts a non-compliant bid, then that becomes a matter between the contractor and the customer. In other words, the fact that AWIL was able to prove that the Sikorsky bid was not compliant became a 'so what?' because of the Double N decision."[12]

Shortly after the announcement was made that AWIL was dropping its lawsuit, Sikorsky announced that it could not deliver the *Cyclone* on time. On 10 January 2008, the Canadian media discovered that the military staff at 12 Wing in Shearwater had been told by Sikorsky that the first new helicopter would not arrive until 2010 or 2011. The pilots, mechanics, and technicians being assembled to conduct trials with the first aircraft were also told that they would not be needed until then. Former Deputy Commander of the Maritime Air Group, Colonel Lee Myrhaugen, explained that new engineering requirements as a result of technological advances could be responsible for the delay and could mean altering the contract. He also concluded that "the likelihood of making the Sea Kings survive [the delay] is extremely limited."[13] Helicopter crews were already getting only a fraction of the flying hours they were once required to have to maintain proficiency.

Chief of the Defence Staff General Rick Hillier gave a speech at the historic Pier 21 in Halifax Harbour that day. He said that he was frustrated by the delay and that Canada needed to shed its reputation of being "world-class at maintaining old equipment."[14] Peter MacKay, the Conservative Minister of Defence, blamed the previous Liberal governments for the delays in the delivery: "It's a tremendous, tremendous disappointment to see once again this vital piece of equipment may be delayed. And it can all go back to a single, solitary decision and a flippant and callous stroke of the pen."[15]

In an e-mail to the Canadian Press, Jacques Gagnon, the communications director in PWGSC, wrote that "we are assessing the implications of what a delay on the delivery of the maritime helicopters will have on the operational requirements of DND and PWGSC is considering all possible options with respect to Sikorsky's default on the timely delivery of the Maritime Helicopters."[16] Part of these implications would have been late fees for breach of contract. But the maximum amount of penalties that the government could impose was fixed. And the RFP indicated that the delivery schedule for all helicopters was tied to delivery of the first one. The consequence of this provision is that additional liquidated damages for late delivery of all subsequent helicopters would not come into effect provided that one is delivered every month after the first delivery – regardless of how far behind schedule it is. Provided that deliveries remain uniformly late, the maximum liquidated damages would be $36 million.

In a further attempt to contain criticism over the revelation that Sikorsky could not deliver its helicopter to the Canadian Forces on time, Dan Ross, the new ADM Mat, responded to an article that I wrote:

> The Cyclone project was contracted prior to the transition to performance-based procurement. Still, the solution is based on the proven civilian version of the Sikorsky Cyclone. The DND and Public Works and Government Services Canada are working with Sikorsky to update the Maritime Helicopter project schedule, and to mitigate, as much as possible, the impact of delays on the operational capabilities of the Canadian Forces.[17]

It is fascinating to note that the process for major military equipment procurement for the CF based on "performance" was being considered as a new trend at DND and PWGSC. This style of procurement has stressed best value and clearly defined objectives and timelines. This obvious business strategy had actually been recommended by the Standing Committee on National Defence and Veterans Affairs (SCONDVA) in 2000. Although the ADM Mat attempted to reassure the public that problems similar to those of the MHP would not occur under his watch, the damage to the MHP and the CF had occurred long before Sikorsky's announcement. And it makes little difference what model the future *Cyclone* is based on if it is not certified and ready to enter service in Canada under the four-year timeline stipulated in the contract.

By May 2008, it became clear that the original contract of 2004 would have to be renegotiated. Not only was Sikorsky unable to deliver the helicopters, but they demanded more money. It was speculated that the company was asking for as much as $500 million extra to upgrade engines, mission suite computers, gearboxes, and rotor blades.[18] Neither side has ever released any details on the discussions. But the meetings took even more time off the clock for the *Sea King* helicopters, which were, by that time, expected to have to operate into the 2012 timeframe. The aircraft's navigation and communication equipment, as well as its radar, were still incompatible with the needs of a modern navy. Supportability studies had to be initiated to address air worthiness, parts obsolescence, and general sustainability. An inquiry into the *Sea King* crash the night of 2 February 2006 off Her Majesty's Canadian Ship (HMCS) *Athabaskan* found that the pilot had not successfully requalified for night flying because there were not enough serviceable helicopters to conduct proper training.[19] Colonel Lee Myrhaugen explained: "Because the equipment is sometimes no longer supportable, the aircraft is bordering on redundancy."[20] His worst fear was that the air force would conclude that the aircraft could no longer support the ships that had been made to carry them.

The Canadian government originally gave the impression that it would take a hard line with the company. PWGSC Minister Michael Fortier asserted that he still expected Sikorsky to meet its contract obligations as signed. He stated: "When the government signs a deal with a supplier for a specific good at price x, that's the price the government should pay for that good."[21] It was also claimed in the media that the government was withholding a $200 million payment for failure to reach essential production milestones.[22] There were even rumours that the government was threatening to cancel the whole contract. This bravado did not last long. Transcripts of the Senate Defence Committee revealed that Lieutenant General Angus Watt, Chief of the Air Staff, involved in the private negotiations with Sikorsky, believed that "the initial indications are that the delay will be a matter of some months to years, but well within what I view to be acceptable."[23] Minister of Defence Peter MacKay was also optimistic about the negotiations. He proclaimed to the press that "we will get an aircraft that is improved from its original contract and the timeframe and the costs between our position and theirs have been reduced significantly."[24] But he admitted that costs had risen: "You get what you pay for, and if we're getting a better aircraft of course, it's in keeping with commensurate fees associated with increased capacity." This statement ignored, of course, the fact that there were never supposed to be major changes to the aircraft or the contract after it was signed – especially changes that would delay delivery and add significant expenses. The minister knew this and put on a brave face to defend the indefensible. MacKay had already declared the attempt to replace the *Sea King* "the worst debacle in Canadian procurement history" and blamed it on Chrétien's decision to cancel the *EH-101* contract in 1993.[25]

On 23 December 2008, an agreement had been reached through arbitration. MacKay, alongside the new Minister of PWGSC, Christian Paradis, announced amendments to the contract with Sikorsky for the MHP. They continued the effort to portray the protracted procurement in the best possible light with the least possible audience. An announcement of this kind made a couple of days before Christmas was no coincidence. Paradis claimed that "these amendments represent a significant accomplishment in getting compliant maritime helicopters with a minimum of delay, while protecting the investment of Canadian taxpayers."[26] MacKay proclaimed that "the Canadian Forces will now receive their first Cyclone helicopter in November 2010, a date that will allow our men and women in uniform to continue their outstanding work." One wonders if he had his fingers crossed for good luck. The necessity of the portrayal of optimism, however, is understandable. Nobody really expected them to admit that their predecessors had simply chosen the wrong aircraft and that everybody had to pay for the mistake but them. The DND web page even labelled the release "Canadian Forces to Receive Helicopter Fleet with Leading Edge Technology" to try to mitigate the fallout. Although the *Cyclone* that is scheduled to arrive in Canada in 2010 will be in a troop transport configuration only and will not be fully capable, the webpage applauded how the first *Cyclone* test flight had been conducted successfully in the United States in November 2008. Nobody mentioned that this was the date that the first complete helicopter was supposed to be delivered to Canada. Finally, the release attempted to educate the public on how such procurements work: "In complex procurements such as this one, it is not uncommon to encounter delays."

The contract amendments have an estimated value of $77 million for the helicopter contract and $40 million for the twenty-year In-Service Support contract. The press release pointed out that this extra cost represented less than 3 percent of the MHP investment of $5 billion – almost as if it did not matter – a non-issue that the public should simply accept as the cost of doing business. It even described the new developments as a saving to the taxpayer. It also revealed that the government had determined that the two-year delivery delay was "largely outside the control of the Contractor." In contract terms, this is called an "excusable delay." It was speculated that Sikorsky argued that the delays were not its fault because the helicopter that was accepted during the bidding phase was not compliant with the DND's requirements.[27] Indeed, if the DND shifted the requirements and asked for changes to be made to the *Cyclone* beyond what was originally agreed to, on the surface this is a justifiable claim by Sikorsky. And the DND had admitted that "Sikorsky is making a number of improvements to the current design of the helicopter to meet the performance requirements specified in the current contract. An improvement being made that was not in the original contract will provide the helicopter with growth potential for the engine and main transmission."[28] But this whole concept of growth potential, for the engine or otherwise, was never supposed to be an issue. The weight growth requirement was actually stated in the MHP Statement of Requirement (SOR) and was initially included in the RFP requirements specification but was later removed at Sikorsky's request. More importantly, any design changes contradicted the agreed on lowest-cost-compliant procurement process used. The DND actually had to assure the Auditor General that there would not be any need to provide additional capability during the service life of the aircraft. With that assurance, the DND evaluated the competing Sikorsky *H-92* and the AWIL *Cormorant* bids without assigning value to capabilities that might have exceeded the specification.[29] None of these details, however, reveal the specifics of Sikorsky's "default," as described by

PWGSC. The refusal to offer any information on the contract dispute provides a way for the government and Sikorsky to simultaneously avoid ownership of the fiasco. The DND and PWGSC had actually entered an information blackout.[30] Requests under the Access to Information Act (AIA) were essentially ignored or took over a year to be processed. As a result of his constant failed attempts to secure information about the MHP, Pugliese proclaimed that "the Maritime Helicopter project office at DND is one of the most secretive organizations at the department and over the years has provided almost zero information about how it is spending 5 billion dollars of taxpayer's money." The DND would not even reveal who was in charge of the project. When there were responses from the DND or PWGSC, they were classic examples of word manipulation and obfuscation. This was a clear example of skirting public accountability. Sikorsky followed its customer's example; in fact, it has made few public statements since winning the award in 2004.

Another of these inexplicable events – on a long list of baffling MHP acquisition decisions – was that the government also decided there would not be an assessment of liquidated damages for non-compliance of the contract even though it stipulated that a penalty of $36 million could be claimed for any failure to deliver the helicopters on time. Former ADM Mat Alan Williams asserted at the time of the original contract that the Canadian citizen could be "certain that we will hold them to account."[31] But by 2009, Williams had retired from the DND. Michael Fortier from PWGSC had also left office. After the first reports that Sikorsky would be late in delivery, MacKay also sounded a warning to the company: "We're going to try to pin them down a little bit further on what the timelines are. There are penalties and clauses that will kick in."[32] But they never did. No reason was given for the decision.

On 12 March 2009, a civilian version of the future *Cyclone* crashed off the coast of Newfoundland and killed fifteen passengers and two crew. They were workers being ferried from St. John's to an offshore oil platform. There was only one survivor. The investigation conducted by the Transportation Safety Board of Canada (TSB) discovered that the primary reason for the crash was that two of the three mounting studs that hold down the oil filter assembly on the main gear box were damaged in flight.[33] This resulted in a rapid loss of oil pressure and failure of the gearbox. The results of the investigation also revealed that Sikorsky was not technically compliant with the advanced standards of the US Federal Aviation Administration (FAA).

Specifically, the aircraft was never compliant with their Federal Aviation Regulations (FAR) Part 29 that required the main gearbox to run for thirty minutes without oil. The regulation is there to give the pilots time to find a suitable emergency landing zone. In this case, it was only ten minutes after the pilots reported oil pressure problems that the helicopter ditched into the ocean after the tail rotor lost power. As Sharon Hobson has written, a May 2004 report by the Joint Aviation Authorities (JAA), the European counterpart to the FAA, revealed that after the gearbox failed the thirty minute loss of lubricant test, it was modified to isolate "the external oil cooling system in the event of a failure, which allows rapid loss of oil from the cooling system." The FAA agreed that this effort provided an effective method of allowing continued operations in such circumstances and gave the company an exemption based on their demonstration that all other possible failures of the main gearbox that could result in rapid oil loss were "extremely remote."

Sikorsky spokesperson Paul Jackson quickly defended their aircraft by saying, "The *S-92* meets the highest safety standards ... Nothing is more important than safety." He then reiterated the typical company supposition that "The *CH-148 Cyclone* will be among the world's most advanced and capable maritime helicopters."[34] Clearly this was an accident and there was no malicious intent on the part of Sikorsky. And it was the FAA that allowed the exemption. But it was not the first such incident. In July 2008, another *S-92* in Australia was forced to make an emergency landing after suffering an identical failure to the one involved in the Newfoundland crash.[35]

Sikorsky is currently being sued in a Court of Common Pleas in Pennsylvania by the family members of fourteen of the deceased and the sole survivor, Robert Decker.[36] Among others, the lawsuit names Sikorsky Aircraft Corporation; its subsidiary, Pennsylvania-based Keystone Helicopter Holdings; and Sikorsky's corporate parent, United Technologies Corporation. The plaintiffs seek unspecified damages.

The lawsuit claims that Sikorsky falsely represented that their *S-92* would be able to fly for thirty minutes after the main gear box lost its oil, which led directly to the fatal misperception by the pilots in question. It reads: "Because of the Defendant's misrepresentations and misstatements, the pilots were not aware that complete loss of operational control was imminent, and therefore, they did not attempt to set the aircraft down immediately on the water while the pilots still maintained control of the helicopter." The lawsuit illustrated how the passengers onboard the *S-92* would have suffered pre-impact fright "with full comprehension of the extent of their peril, and further suffering the effects and pain of injuries sustained in the crash and in their drowning." Robert Decker suffered multiple fractures and hypothermia, and he nearly drowned while awaiting rescue. The lawsuit states that the *S-92* is a flawed aircraft and that the Sikorsky corporate culture valued marketing above engineering and covered up its deficiencies to avoid costly redesigns.

Indeed, it is easy to see how a misconception could occur on the part of the pilots. The FAR 29 requirements demanded a run-dry capability for thirty minutes, and the *S-92* was certified under them. Nobody would know about the exemption unless Sikorsky explained the danger in full to their clients. The lawsuit claims that this explanation was never given. The lawsuit also blames Sikorsky for choosing titanium, which can be a brittle metal and can, without detectable signs, fail under the fatigue of constant vibration common in a helicopter. The document asserts that the issues regarding the titanium, and the main gearbox itself, were highlighted after the Australian incident and clearly established the risk level as higher than "remote."[37]

Although the lawsuit admits that Sikorsky issued a safety advisory regarding the titanium studs, maintenance crews were only told to check the visual portions of the studs during the normally scheduled replacement of the main gear box oil filter. Sikorsky recommended that the studs be replaced within one year or 1,250 hours, whichever came first. There was never a clear admission that the danger was immediate or that the helicopter could not run dry for thirty minutes. The FAA issued an emergency airworthiness directive for the removal of the titanium studs and their replacement with steel after the Newfoundland crash. The cost to do so on the $15-million helicopters was $480 per aircraft. The FAA also ordered Sikorsky to update their *S-92* flight manual to accurately reflect the dangers of losing oil from the main gearbox and to make it clear that the pilots would need to land immediately in such a situation.

This crash is a crisis multiplier for the MHP. The *Cyclone* contract also includes the requirement for a thirty-minute run-dry capability and clearly loss of oil within this design is beyond problematic. Annie Arcand, a spokesperson from DND stated on 8 April 2009: "The run-dry capability requirement for the CH-148 gearboxes, as stipulated in the solicitation document, is included in the maritime helicopter contract," she said. "All requirements are addressed during each major review during the design process."[38] These reviews include the System Requirement Review, Preliminary Design Review, and Critical Design Review. But Jackson later admitted that "we are designing a system to meet the programme requirements and will test and enhance it as necessary." Hobson correctly suggests that this statement may mean that a design for the run-dry capability was not approved during the Critical Design Review and was not included in the first test aircraft. MacKay reverted to his hardline stance with a statement claiming the Canadian government would not accept any helicopter that was not compliant with the run-dry requirement.[39] The *EH-101,* quite predictably, had already been run-dry certified for years.[40]

On 18 June 2009 the TSB reported that the gear box was not the only thing that failed on the aircraft – the flotation system was activated, but the compressed gas to inflate the collars did not deploy even though the activation switch was found in the armed position after recovery. As a result, the aircraft sank to the bottom of the ocean. At the time of writing, it was still unclear why the system failed, but it was assumed that it may have had something to do with the speed at which the aircraft hit the water.[41]

If the first *Cyclone* does arrive in 2010 – and there is no reason to believe that it will – this arrival will only allow for operational testing and training. The CF will have to wait for enough fully capable helicopters to actually replace *Sea King* operations. The complete fleet is not due until 2013 – fifty years after the first *Sea King* came to Canada. The reality is that the government was forced to forgive Sikorsky's offence. PWGSC and the DND accepted the amendments put forward. The department officials realized that there was no way to compel performance; if they went public and threatened cancellation – in itself political suicide in light of the 1993 cancellation – it would be revealed that the DND and PWGSC had awarded a multi-billion-dollar contract to a non-compliant company. Such a disclosure would make them vulnerable to legal battles with Sikorsky. The company could easily claim that the government had requested changes to the overall requirement and that, therefore, it was not Sikorsky's fault. This scenario would also open up volumes of secret documents that could prove the government's mismanagement of the project from the beginning. In this case, the government had ignored the rule of *caveat emptor* and was left holding the rather sizable bag as a result.

The final budget for the MHP is unknown. The final delivery date is unknown. It is even likely that the $117 million will not be enough to make the current aircraft compliant with Canada's needs. There has been no explanation of which design changes that money is for, and more amendments will almost certainly be needed. If the changes revolve primarily around an engine upgrade because the aircraft is heavier than originally expected, this work could result in a major escalation of labour in other areas that will require extensive testing. This situation could also affect the actions and schedules of each independent Sikorsky subcontractor and thereby increase overall lead times. Further delays could continue to come from the subcontractors themselves. GDC's mission system, for example, is not ready to integrate into the aircraft even though those were also due in 2008. Company spokesman Paul Jackson has admitted that the encrypted communications system will not be ready in time for the new 2010 delivery.[42]

Everything that was specifically outlined that would put the MHP in jeopardy has happened. And nobody will be held accountable. Most of the major players responsible for the scores of questionable decisions regarding the

MHP are either retired or outside the scope of punishment. It no longer matters that the government selected a piece of kit that did not meet its stated performance requirements. That is yesterday's news and will likely remain the most overlooked defence story for years to come. Inexplicably, even the Auditor General has refused to investigate the acquisition.[43] This is defence procurement in Canada.

At the time of writing, the *Cyclone* had yet to fly in Canada. It did not fully exist in a military format. It was still a developmental aircraft, and the DND and PWGSC knew this when the contract was awarded. AWIL certainly knew it, which was made clear by their warnings to the government. And Sikorsky knew it too. In fact, the United Technologies Corporation admitted as late as 2008 in its Annual Report that: "The (CH-148) is being developed under a fixed-price contract that provides for the development, production, and 20-year logistical support of 28 helicopters. This is the largest and most expansive fixed-price development contract in Sikorsky's history."[44] This procurement occurred even though the importance of buying off the shelf and avoiding the risks of using untested equipment had been repeated countless times by analysts and military professionals along the way. The reality may be that the government agreed from the beginning to take deliveries beyond 2008. Although the government departments must share the blame for awarding such a contract, the real losers are the Canadian taxpayers and those who risk their lives in the Canadian Forces.

As a result of their experience with the *CF-18* procurement, the military and the government agreed that they did not get good value from a first production run. The contract with EHI for the NSA, therefore, was signed only after the British had committed to the *EH-101*. After the NSA was cancelled and the new MHP was initiated, this policy was reversed. This is not to say that the final version of the *Cyclone* will not be a capable aircraft for the Canadian Forces. This is completely unknown. It is simply an untested aircraft, and the manner in which it was selected has not inspired trust. The process, from the first SOR to the final RFP, was manipulated to allow weaker contenders to enter the competition. And the final evaluation allowed for one of them to win. AWIL was deemed non-compliant, but in reality it was a politically unacceptable choice. The lower price tag of the *Cyclone* allowed the government to choose an alternative model and therefore avoid the embarrassment and obvious criticism that would have surfaced had it chosen the *EH-101* again.

Conclusion

> In competition with other departments which require larger budgets to provide better public welfare and public services the needs of [Canadian] defence tend to be regarded as wasteful and negative. The vigor of this opposition to defence spending fluctuates with the international situation but is always present.
>
> – Douglas Bland

> Defence acquisition by its nature is a complex, expensive and technology intensive business. Procurements are uncertain, both in terms of schedule and requirements, with a small pool of suppliers dependent on winning the relatively few contracts to survive. The process is financed by a government that understands little of the military requirement, must divert dollars from other pressing needs to support defence procurement, and therefore seeks to achieve many other non-defence objectives simultaneously from the same dollars. It is inherently a very risky process, overseen by a government that is extremely risk averse. That it delivers anything at all should be quite surprising.
>
> – Pierre Lagueux, former Assistant Deputy Minister (Materiel) 1996-99

Politics has steered military procurement in Canada throughout the country's history. The cancellation of the New Shipborne Aircraft (NSA)/New Search and Rescue Helicopter (NSH) project in 1993 and the choice of the Sikorsky *Cyclone* in 2004 comprise the ultimate case study. The NSA/NSH illustrated how defence contracts cannot be sustained indefinitely in a country where the majority of citizens do not support investment in defence. And when it came time to try again to replace the aircraft, the government intervened in the procurement process and sculpted it in such a way as to prevent purchasing a helicopter that was politically unacceptable. The government could not choose an aircraft that it had already refused to purchase in a prior contract and paid hundreds of millions of dollars to cancel. The *Cormorant* never had a chance. Such a choice was impossible to justify to Canadian citizens. In both cases, it made little difference whether the proposed equipment was ideal for the needs of the Canadian Forces. This is not how weapons and equipment are acquired in this country. The political atmosphere must be favourable to the acquisition and remain that way throughout the procurement process. Timing is everything. If political support wanes along the way, the acquisition is in jeopardy. It also makes little difference whether a contract has already been signed. And because DND procurement officials are aware of these certainties in their business, they too must play the game. They have not been free to communicate what they believe the military needs for fear of their choices being rejected outright by their political masters. The purchase of materiel must conform to a strategic choice for whichever political party is in power and must not embarrass the government later. As a result, rarely has equipment been planned for and delivered within the same government term. Perhaps most importantly, it is the military that suffers in the field of battle as ministers and officials from the Prime Minister's Office (PMO), Public Works and Government Services Canada (PWGSC), the Department of National Defence (DND), and an array of other departments engage in the political dance that is procurement in this country.

Procurement in Canada has had a dilatory history that is not unique to the replacement programs for the *Sea King* – less extreme, perhaps, but not unique. The Avro *Arrow* is obviously the second most famous example of a failed procurement due to cost overruns and delays. The procurement of the *Mackenzie* (*Repeat Restigouche*) class of ships was another example of where the shift in political leadership could have ended the replacement of Anti-Submarine Warfare (ASW) destroyers necessary for the Royal Canadian Navy (RCN) to carry out its North Atlantic Treaty Organization (NATO) mandate. In this case, the new Diefenbaker government recognized the necessity of the project and carried it through. It could have just as easily been cancelled. The navy had taken too long to agree on a ship design for the replacements, and this indecision made the project vulnerable after the election. Diefenbaker agreed to go ahead with the program partly because he saw the political advantage of building ships in Canada. When he was defeated by Lester Pearson in 1963, the situation was reversed. The RCN had been working on the General Purpose Frigate (GPF) to replace its aging vessels and carry out its international commitments. But because there had not been any domestic design of ships in Canada since the *Mackenzie* class, there was no industrial memory left to create a modern vessel. There was a lack of skilled engineers and draftsmen. The Department of Defence Production had a similar problem and did not possess the technical staff or knowledge to assess risk and control escalating costs. As the price tag nearly doubled, the new Liberal government took power and quickly cancelled the program. It had taken too long and was too expensive due to industrial inexperience and lack of skilled labour. However, shipbuilding has been the most successful area overall regarding Canadian design and production of equipment for the Canadian Forces (CF). Other examples of the dominance of Canadian political considerations over military capability in the field of weapons and equipment procurement have been covered in the Introduction.

Procurement of the *Sea King* itself followed the classic pattern of indecision, delay, and unforeseen expense that has always come with reliance on foreign equipment and Canadian attempts to modify that kit to fit national requirements. Canadian industry has rarely designed and produced its materiel for war. And when it has tried, it has often failed. This failure has been due partly to massive overspending, an inability to deliver on time, or the production of a piece of kit that is ineffective in comparison with foreign models. Indeed, the 1964 Defence White Paper acknowledged that Canada would continue to look to foreign sources for the majority of its military materiel. The delays involved in the *Sea King* acquisition, therefore, were somewhat to be expected. The ship conversions, training, and actual delivery were all complicated by the need to consult and pay the United States regarding each matter.

Canadian industry was not considered capable of producing a modern ASW helicopter at a reasonable price; however, the alterations to the *Sea King* were completely in line with national military considerations and not politically driven. The modifications to the helicopter could not be done anywhere else because the technology involved existed only in the minds of RCN officials, and it was integral to their vision of shipborne helicopter operations in the 1950s. Although only a small piece of military hardware, the *Beartrap* was one of Canada's most original contributions to the field of international ASW, alongside Variable Depth Sonar (VDS). Canada had only one aircraft carrier, and it was correctly believed by naval officials that the country could increase its ability to carry out its NATO ASW duties if helicopters could fly from smaller ships. It was also realized that helicopters could make up for the lack of Canadian surface escorts. The extensive delays in certification of the *Sea King* fleet in Canada were due to the necessary trial and error that came with the design and production of a unique mechanical innovation. Indeed, the Canadian *Beartrap* was a new innovation, and faults in its early production were tolerated since there was no technological alternative. By the late 1960s, the Canadian Navy was recognized as the leader in the field of shipborne helicopter operations. After the *Bonaventure* was decommissioned in 1969 and fixed-wing aircraft operations were ended at sea, the *Sea Kings* became integral to ASW in Canada.

By the 1980s, the *Sea King* was reaching the end of its operational life and had to be replaced. Although the *EH-101* had been selected for this purpose in 1987, by 1992 a final contract still had not been signed. This was due not to the necessity of designing or producing the helicopter but to choices made by the government. The Mulroney government had decided to link defence acquisitions with industrial offsets for Canada and demanded that any major equipment purchase provide for the involvement of Canadian industry. Procurement of the *Aurora, Leopard I, CF-18,* and Canadian Patrol Frigate (CPF) projects had made a direct connection between Canadian economic development and the purchase of military equipment. By the time the government decided to replace the *Sea Kings,* the investment in regional industry dominated military acquisitions. This policy was upheld and expanded in the 1987 Defence White Paper. As Gordon Davis has written, "the 1987 Defence White Paper articulated a defence policy that recognized the Cold War realities of the day and committed the government to re-equip the armed forces to enable them to carry out their missions. However, in so doing it transformed defence policy into an industrial development programme."[1] European Helicopter Industries (EHI) had decided, therefore, that it would install a Canadian-built electronic mission suite in the *EH-101* airframe, even though there was one readily available that was to be used by the British and Italian navies. The multitude of subsequent contracts and subcontracts with Canadian industry became even more unwieldy when the government decided to integrate the NSH program with that of the NSA in 1990. This decision meant that EHI had to go back to the drawing board and redefine how it was to implement the government policy of stimulating Canadian industry for both programs. The concept of including what was eventually speculated to be approximately 400 companies in the purchase of a military helicopter created a myriad of bottlenecks that was increasingly devoid of any substantial risk assessment by the NSA project office.

The growing number of ministries and committees within the government and DND created competing priorities and slowed the procurement process considerably as each of their concerns had to be taken into account before moving forward. These lulls did not threaten the project directly. Nor did they weaken the potential of the final aircraft that was expected; an *EH-101* with Canadian-designed electronics would have been a great success for naval operations in Canada. There is a multitude of advantages to modifying equipment to suit domestic needs. Canadian history is full of examples of how the military was hampered by equipment that had been designed for some other national military.[2] There are also clear national economic and development advantages to the stimulation of domestic industry, especially in high-technology sectors such as aerospace. IRBs, if carried out properly, are a worthwhile lever to achieve broad national objectives through weapons acquisition; they demonstrate that Canada will act strategically to nurture its industrial capabilities. Although it was never stated sufficiently by the NSA project office or the DND that the program was the best way to create jobs, it was to do so while procuring a piece of multi-role equipment considered necessary to maintain military capability. And the jobs being created were highly technical; they would help to develop Canadian industry and create spin-off employment and participation in the international marketplace. But by the 1990s, the Canadian procurement cycle had become too cumbrous and bureaucratic, and nationwide industrial involvement created extensive delays. The *Sea King* fleet needed urgent replacement, and the government did not have the luxury of the time needed to carry out this type of protracted acquisition.

For most nations, military purchases represent the largest proportion of government discretionary capital spent on industrial goods.[3] How it is spent, therefore, is of vast importance. Part of this spending usually entails a domestic industrial defence base, which has been described by Colonel W.N. Nelson as:

> The capability of Canadian industry to support Canada's defence undertakings, in times of both peace and war. Defence Industrial Base includes the infrastructure and personnel required to produce modern weapons systems, spare parts for weapons systems, munitions, and supplies. It also includes the infrastructure and personnel to perform equipment repairs and overhauls, as well as to conduct research and development.[4]

Nelson continued by asking, "what about Canada's current and future military undertakings? Does it matter whether Canada relies on foreign sources of supply to fulfil these undertakings? Are there reasons for having a stronger defence industry in time of peace?" Greig Stewart has put forward the advantages of producing military equipment domestically:

> When Canada spends money in Canada to design, develop, and manufacture high technology products ... the unit cost of such items is not of overriding importance. The money is spent in Canada to provide jobs for Canadian workers who in turn pay taxes and buy goods and services, helping to strengthen other Canadian companies and the economy as a whole. Design, research, and development are investments in the future, raising the level of Canadian technology and lowering our reliance on foreign technology and expertise.[5]

But others have pointed out the economic difficulties of such an attitude in Canada. John Treddenick has astutely asked,

> How should one treat a particular defence production activity which generates considerable domestic employment but produces a weapons system at greater cost than an equivalent [product]? On the one hand it contributes to the goal of full employment, yet on the other, it reduces the military capability which could be achieved from a limited defence budget. In the absence of a higher order criterion, such tradeoffs are difficult to disentangle and economic significance difficult to assess.[6]

It is not an obvious decision to design and build weapons and equipment in Canada. But it is equally clear that there are many disadvantages for a nation that must rely on foreign sources, especially if it wants to make domestic alterations.

Dan Middlemiss is one of the few authors on defence procurement in Canada and has argued that it is a vital component of defence policy: "It is what puts the 'arms' into the armed forces and because of the many (sometimes very large) contracts and jobs involved, it is also 'big business' in Canada."[7] But as Treddenick revealed in 1988, the overall impact of defence procurement in the Canadian economy was minimal: "Total defence production accounts for considerably less than 1 percent of both gross domestic product (GDP) and total employment."[8] Despite this fact, Middlemiss's point still holds. There are still large contracts to be had. And in Canada, the issue of economic offsets, regional development, and employment can become more important than the military operational requirement. As Craig Stone has concluded, "despite the relatively small impact to the overall economy, the dominance of domestic economic and political considerations in Canadian defence capital spending, to the relative neglect of security or strategic military factors, is the normal defence climate in Canada."[9]

The defence procurement objectives of the NSA/NSH involved maximum effectiveness, modern technology, maximum interoperability and flexibility, and training and maintenance, all at the lowest cost. This list conflicted with the non-military objectives of enhancing Canada's industrial technology base across the country and increasing Canadian exports. The 1992 report of the Auditor General noted that the 1985 Ministerial Task Force on Program Review Concerning Government Procurement concluded that the DND was concerned that operational needs were being jeopardized by ad hoc, third-party decisions of other government departments.[10]

Further delays in the NSA project resulted from another Canadian procurement strategy at the time. It was determined to avoid the purchase of the first *EH-101* models coming off the production line. The DND had already encountered longevity problems with acquisition of the first run of *CF-18* jet fighters. The department decided to wait until the British had signed their contract with EHI. The final contract to deliver the *EH-101* was not signed until 8 October 1992 with a total project cost of $4.4 billion. It was not solely the fault of the project office that it took so long. The government's procurement policy had made the delays inevitable. The consequences for the project, however, were not readily apparent until 1992, when the program suddenly became a focal point of attack in an election campaign.

The earlier procurement of the *Sea King* was difficult because the roles of the naval helicopter were not yet clearly determined by the naval staff in the 1950s. This is understandable since the discussion centred on what was to be a new platform of ship and helicopter. It was also unclear which service was going to be responsible for the aircraft once they were procured. But by the time the *Sea King* needed replacement, military officials, including the naval staff, Chief of the Defence Staff, and Minister of National Defence, were clear on what roles helicopters would be fulfilling and who was responsible for them. In addition to the classic ASW capability, the security environment of the 1990s demanded that the aircraft also be used in the maintenance of Canada's sovereignty and security through active participation in drug interdiction, environmental monitoring, fishery protection, Search and Rescue (SAR), international peacekeeping and humanitarian aid, control of maritime approaches, and response to international contingency operations, including support to land force operations. These issues had been highlighted in the 1987 Defence White Paper.

The early 1990s saw the roles of the *Sea King* expand during the Gulf War, Somalia, Haiti, and the former Yugoslavia and made it even more apparent that a modern naval helicopter was needed to keep pace in the global theatres where Canada would be asked to operate. The procurement officers in charge at National Defence Headquarters (NDHQ) quickly understood that the end of the Cold War would not mean a decrease in CF operational tempo and that the NSA/NSH project had to be carried through. The flexibility of the *EH-101* aircraft made it perfect for Canada's needs. Although the differences in price for all the possible choices were relatively small, the level of capability of the *EH-101* over its competitors was enormous. Although the DND and the NSA/NSH project office occasionally shifted their emphasis toward the newer roles over that of traditional ASW to make the project more politically acceptable to Canadians, these functions were all integral to the procurement. But because of a powerful misinformation campaign mounted by the official Liberal opposition, most Canadian citizens did not understand this until it was too late.

It was clear that there would be an election in 1993, and many issues other than helicopters became important. Many voters likely did not even know what the NSA/NSH project was. Of those who were aware of it, many refused to support it because of how it was portrayed by the Liberal opposition. The NSA in particular was depicted as an extravagant expense, and its value in terms of capability became extraneous. The Liberal stance was easy to believe because the Canadian people had no faith in the Conservative government by 1993. When the Conservatives came to office in 1984, the federal deficit was at an unprecedented high. Notwithstanding the government's pledges to reduce it, the deficit had only grown by 1993. In an attempt to restore the fiscal balance, Mulroney had brought in the highly unpopular Goods and Services Tax (GST). The Conservative government was behind in the polls for most of 1992. In truth, there is little that it could have done to save the NSA/NSH project. The 1993 election saw the Conservative Party suffer the worst political defeat in Canadian history.

The first prong of the Liberal attack was to use the figure of $5.8 billion instead of the amount that the contract was signed for – $4.4 billion. The former figure was an estimation of what it would cost if inflation was factored in over thirteen years. This tactic would not have been unscrupulous had the Liberals not claimed that the costs had risen from $4.4 billion to $5.8 billion after the contract was signed. The true misconception among the public, however, was that the $5.8 billion was to be spent that year. During a time of recession, any citizen would balk at such a figure. It would have been over half of the total defence budget. The reality, however, was that this amount was to be spread over the full thirteen years of the project; there would have been an approximate investment of $200-$300 million the first year or approximately 3-4 percent of the defence budget each year after that. The Liberal tactic, however, was highly effective as defence expenditures have garnered little support throughout Canadian history due to the sensitivity of Canadian citizens regarding military commitments. All parties involved in the NSA/NSH were aware of this. When describing the abilities of the *EH-101* to track and kill enemy submarines, for example, NSA project officers used the word *prosecute* instead of *kill* or *destroy* to placate the public.[11]

The second Liberal prong of attack on the NSA was much simpler but just as effective. The Liberals claimed that, because the Cold War was over, there was no need for a naval helicopter. They talked of its attack capability and refused to acknowledge any other use for it. The pervasive fear of Soviet expansionism and technological capability in the 1960s made it a sound investment in the minds of Canadians. The political leaders could carry it out without reproach. But by 1993, with the perception of threat removed, it became much more difficult to sell investment in a platform that had a traditional role based on ASW. After the collapse of the Soviet Union, there was a pervasive belief that global stability would return and that there was no threat from Soviet submarines. There was, by that logic, no need for an ASW helicopter. The reality was that the *EH-101* was a multi-purpose, flexible aircraft. It was equipped to conduct four naval missions equally well: ASW; anti-surface warfare (ASUW); utility transport and support; and SAR. This multi-purpose role was never understood by the public. Moreover, Canadians were unaware that the operational tempo of the CF in the 1990s was increasing. The people were optimistic. In a time of recession, the political environment was closed to major defence expenditures.

The Conservative government and DND did try to explain to the Canadian people the uses of a flexible helicopter in the post-Cold War world. This campaign even extended to the prime contractors of the NSA/NSH project. Paul Manson, President of Paramax at the time, toured the country to explain to the editorial boards of every major newspaper the advantages of the project. There was, indeed, a public information campaign. It simply did not achieve the necessary results. Defence procurement is a difficult sell in Canada, and it was easier to believe the other side. The investment in high-tech industry and the jobs to be created across the country did not comprise a good enough reason to carry out the contract. Although it was considered a good idea to invest in our nation's industry, this idea did not extend to the military sector.

The public had to be educated on the instability that would occur after the Cold War and on the roles that naval helicopters had acquired throughout their operational lifetimes, expanded functions that continued to make them necessary. The results of the 1993 election, in part, proved that this campaign was not a success. The Liberal claim that the helicopter was a waste of money and an unnecessary relic of the Cold War was repeated so often that it came to be seen as fact. Although the other issues of the election, such as the recession, were more important overall, the *EH-101* was a symbol of extravagance in a time of fiscal restraint. Moreover, the people of Canada did not understand the complexities of the military procurement system. It was believed that the contract would be cancelled with few repercussions and that a helicopter could be bought quickly at a later date when prosperity returned.

The Conservative government was already weakened and disorganized by 1993, and defence of the NSA/NSH project was generally mishandled. Prime Minister Kim Campbell demonstrated a lack of resolve when she gave in to pressure to reduce the final number of helicopters to be procured. As Minister of National Defence, she had already claimed that this reduction was not possible. The amount of $1 billion in savings as a result of cutting seven NSA was also never believed. The credibility of the NSA/NSH project, and the Conservative Party, suffered as a result.

It is fascinating to note that Quebec opposition to the cancellation was muted until after it was complete. It was speculated that the province did not want any criticism to hurt a francophone's chance of becoming the next Prime Minister, even though it stood to lose the most from the failed industrial package of the contract. After Chrétien assumed office, of course, it was too late for complaints. Although those at NDHQ had always been vocal about the necessity of completion of the project, there was also a belief that the election would not be the final word on the matter. Some simply could not believe that this election promise would be kept.

The Canadian people were unaware of what the NSA/NSH cancellation would cost them later, both literally and figuratively. The tax dollars that have gone to the hours worked on a replacement since the 1970s are incalculable. The lesson is clear: time was a factor in the procurement of a naval helicopter for Canada due to its political sensitivity. And the argument that the Cold War was over and that peace and stability would soon follow was simply too tempting for Canadian citizens. As Colonel (retired) John Cody, former Commanding Officer of 423 Helicopter Squadron in Shearwater, explained, "the process had become so skewed with political correctness and 'process,' by the time we got the project to the table it just took far too long for it not to change as it went along."[12] After the cancellation, the new Minister of Defence, David Collenette, visited the base at Shearwater to meet with senior navy and air force personnel to attempt to ameliorate the fallout within the community from the cancellation of the *Sea King* Replacement Program. Colonel Cody recounted that "I remember the day like it was yesterday. He actually put his arm around my shoulder over in the maintenance hangar as he could see that keeping my chin up constantly in front of the troops was taking its toll on me and he said: 'don't worry, we'll get you your new helicopter by 2000.' I remarked that I would hold him to that."[13] Cody continued:

> It was later after the February budget of 94 that I learned the MND [Minister of National Defence] didn't really give a hoot for us, as a few comments from him personally and dealings with his junior staff in particular showed. After the crash of our Sea King in Saint John in April of 1994 I didn't receive one single phone call from the MND expressing condolences. What I did receive a few days later was a call from somewhere in Ottawa (the PMO, the MND's office, I have no idea: lots of voices in the background) early one morning from an anonymous person who was concerned only about restricting the political fallout as a result of the cancellation of the Sea King replacement program and the subsequent crash of the Sea King which killed two of my good friends. The call woke my wife and I up at approx 1 am. I cannot recall exactly what was said but I do recall clearly and unequivocally telling whoever it was that I was not going to play those games and to go pound sand, and I hung up on them. That effectively ended my career and I resigned several months later.

The year after the NSA project was cancelled it was revealed in the 1994 Defence White Paper that the requirements for a naval helicopter had not diminished at all; the strategic environment still demanded that the *Sea Kings* be replaced, and the paper identified the new helicopters as vital to maintenance of the Canadian Forces core combat capability. Moreover, the incidences of crashes, deaths, near misses, ditchings, and emergency landings demonstrated to the public that, in addition to the *Sea King* not being operationally effective, its air worthiness was also in question. New helicopters were a necessity, and this meant that it was time to start over again.

To justify cancellation of the NSA/NSH project, the government put pressure on senior defence officials responsible for the procurement to create a new Statement of Requirement (SOR) with a reduced capability. And this is where politicization of the Maritime Helicopter Project (MHP) really started. Rear Admiral G.L. Garnett had to deal with the reality that, if a new SOR was returned with Cadillac-like requirements, another replacement project could fail. The cost and capability, therefore, were cut in order to sell the project to the government. It was to be an off-the-shelf procurement as much as possible. The MHP planning proposal was approved on 25 April 1994, and the MHP was then registered in the Defence Capital Program. The project seemed to enter stasis at this point as the debate continued on the reduced SOR. In this regard, DND officials failed to deliver a complete SOR on what the forces required. As Admiral Garnett had made clear, the Prime Minister and the MND demanded a SOR with Chevrolet-like characteristics. For over six years, the department worked on the project without seeking formal government direction. Trepidation among defence officials allowed the Liberals to successfully put the project on hold. But knowing the political climate, they believed that they had no choice. They knew that they could not have a SOR that would lead them, once again, to the *EH-101*. Any aircraft was better than no aircraft.

Those against the politically acceptable SOR insisted that it be a document that described what the military needed to conduct its operations, not what military officials thought their civilian masters wanted to read. The requirements of the aircraft, such as carrying capacity, endurance, and operating conditions, specifically the reduction of the international standard atmosphere (ISA) +20°C requirement, were being pared down to levels similar to those of the existing *Sea Kings*. The protests by military officials continued as they realized that, after the SOR was complete, the project office was lowering some of the original capabilities in it within the related document of Requirement Specifications (RS). As they saw it, if the SOR set the "minimum operational requirements," it made little sense to continue reducing them in the RS. Changes to the RS had become the largest point of contention. As Major Barrett had sarcastically written, "you may add data [in the RS] where we forgot it unless we decide later that we didn't forget it."[14] The changes created a theme of bias in the competition as more capable companies saw these changes as a way to let inferior aircraft compete.

Once the SOR was done, the government set out how the helicopter would be procured. With the Canadian Search and Rescue Helicopter (CSH), politicians had failed to get involved early enough to influence the competition; after the Request for Proposal (RFP) was issued, it became clear that the *EH-101* was the best helicopter for Canada. Despite trying to circumvent the rules of the competition in several ways, the government was forced to buy a version of the same helicopter that it had already rejected at a cost of hundreds of millions of dollars. The resulting embarrassment to Chrétien and his government meant that Liberal officials were determined not to let this happen again. In its final form, after much sculpting, the procurement strategy chosen by the government was unprecedented for an acquisition of that type. Once again there would be a lack of uniformity in Canadian procurement, and the process would evolve ad hoc. The strategy included splitting the acquisition into two contracts for the airframe and mission system; using a lowest-cost-compliant matrix; and, in calculating costs, taking no account of the extra investment involved in operating two different fleets of aircraft. The cost saving of using the same helicopter for Canada's naval role was the reason the government had decided to combine the NSA and NSH projects in the first place. Despite the immediate need for new helicopters, the procurement evaluation also ignored the value of signing with a company that had a certified helicopter ready to roll off the production lines. These realities did not have to be the defining factors in the competition, but the refusal to even look at them created suspicion that the procurement strategy was designed expressly to preclude a version of the *EH-101* from winning the MHP contract. The procurement strategy also extended the timeline for delivery from 2005 to 2008. Almost a hundred million dollars were spent to keep the *Sea Kings* in the air during the interim; however, no amount of money could make them competent in modern operations.

The government directed the procurement to award the contract to the company that fulfilled the minimum requirements with the lowest price. And the military officials at the DND went along with the political direction because they knew that it was the only way to get helicopters. The problem was that there was no actual accountability placed on the companies to prove that they could fulfill these requirements – specifically the ability to deliver the helicopters on time. If a company such as Sikorsky misconstrued what it could do and for how much, the government would not know until it was too late. And after a contract is signed in this country, according to the 2007 Supreme Court decision in the case of *Double N,* it makes no difference whether the company can actually deliver on its promise. The losing bidder also has no legal recourse.

The procurement philosophy of the government regarding the MHP had always matched that of Assistant Deputy Minister (Materiel)(ADM Mat) Alan

Williams. His compliance with the government's direction of the project was highlighted in a book that he wrote in 2006: "The behavioural implications for ministers are quite clear. Since they can no longer interfere in the procurement process once it has started, their only opportunity to do so comes prior to its formal commencement ... It must be emphasized that there is nothing wrong with ministerial involvement prior to the beginning of the process." The logic is that, in the Canadian system, it is not considered "interference" for politicians who are ultimately responsible for the operation of the DND and PWGSC to express a preference for one evaluation methodology over another. Williams went on to say, however, that "the problem is when ministers try to distort the form the procurement process will take for political purposes."[15] So whether one calls it interference or not, the MHP procurement strategy given to the DND by the government, before the RFP was issued, falls within the "problem" that Williams outlined. Ministers *were* distorting the process for political purposes. The strategy chosen was completely unconventional and against common business practice. Williams admitted at the time that it increased the level of risk in the project. It also went against the Minister of PWGSC's own rules and Treasury Board policies regarding the procurement of major equipment. Acquiring "best value" for the expenditure of tax dollars was expressly mandated. It also went against what many within the DND had recommended. The mistake of using the split procurement strategy was later admitted and changed midway through the process back to a single contract for the sake of efficiency. This change, of course, delayed the procurement even further. Williams also admitted that "at times they [members of cabinet] have delayed the process by interfering with the marketplace in an attempt to influence the list of respondents to a request for proposals." Again, this was done with the MHP, and it was done before the release of the RFP. Politicians ignored the advice of military professionals and tendering agents to influence the outcome of a procurement before it started. This may not be called "interference" in our system, but, as in the case of the MHP, it certainly had the potential to be biased and unethical.

One more inconsistency regarding Williams's philosophy on political interference will suffice. In discussing the attempts to replace the aging *Iltis* utility vehicle for the army, Williams explained that,

> In an effort to reduce cost and risk, the vehicles had to be "non-developmental," meaning already in use by another army so it could be bought for a predictable cost off-the-shelf. And the army was supposed to receive its new vehicles by 1999. So far so good. Unfortunately, politics then began to interfere. Instead of sticking to this strategy, Minister of Defence Art Eggleton was persuaded by officials from a British Columbia firm called Western Star to distort the process. Western Star had previously provided the military

> with a truck, the Light Support Vehicle Wheeled, and was now offering a new truck they had just built. In February 1999, 16 months after Treasury Board approved the procurement strategy but before the process had formally started, the strategy was changed to allow "developmental vehicles" into the competition. Sixteen months of unnecessary delays. The irony or perhaps tragedy in this case was that Freightliner, the owners of Western Star, announced in 2001 that Western Star would not bid on the LUVW [Light Utility Vehicle Wheeled] contract but would instead be closed down.
>
> More than three years after Treasury Board approved the initial procurement strategy, contracts were finally signed, in October 2002 for General Motors Canada's commercial version (Silverado) with deliveries scheduled between October 2003 and August 2004, and in October 2003 for the Mercedes Benz Canada's SMP [Standard Military Pattern] version with deliveries scheduled between March 2004 and August 2005. Five years needlessly lost due to political interference. And for most of that time Canadian troops were deployed in the Afghan combat zone with vehicles known to be inadequately protected.[16]

This is completely accurate. It is also exactly what happened with the MHP regarding the political choice of including "developmental vehicles." But Williams has always denied that there was any political interference in the MHP. Both examples prove that the procurement process in Canada is managed very carefully by politicians to ensure that the government is purchasing military equipment that is politically acceptable – not necessarily what the military needs.

In 2004, Williams explained that the government had chosen Sikorsky because the company had fulfilled all of the DND's requirements at the lowest price. This had always been the way he calculated best value. Twelve years after the first contract to replace the *Sea Kings* was signed, another attempt was made with Sikorsky to deliver new helicopters. After the evaluation process was explained to AgustaWestland International Limited (AWIL), it filed a claim against the government citing errors in the evaluation and political favouritism. Most importantly, it asserted that Sikorsky was never able to comply with the MHP deadline and therefore should have been found non-compliant from the beginning. AWIL explained that the third-party fairness monitor used had close ties with members at PWGSC and that the most senior officer in the MHP office went to work for the company responsible for the missions system of the *Cyclone,* General Dynamics Canada (GDC), before he was even retired from the DND. Although the history of the *Sea King* Replacement Program and its ultimate failure in 1993 was momentous in itself, it also had the potential to affect the Conservative government led by Stephen Harper as AWIL was suing for $1 billion. As the court case proceeded, the first signs of delays for the MHP arose when workers at Sikorsky walked off the job. As a result of this strike, Sikorsky was granted a six-week excusable delay on delivery of the first *Cyclone* to Canada.

On 26 November 2007, AWIL settled its lawsuit with the Canadian government out of court. There was no monetary settlement. As if on cue, Sikorsky then revealed to those within the air force who were waiting for the helicopters that delivery would be up to thirty months late. Although AWIL had a strong case in challenging the procurement process of the MHP, it chose instead to drop the lawsuit and concentrate on future business. The reality, however, was that the Supreme Court ruling in the case of *Double N v. Edmonton (City)* set a precedent that allowed for faulty bids to be given with no recourse for the other competitors. Sikorsky, therefore, will not be held accountable for its delayed deliveries. No liquidated damages for non-delivery were assessed, likely due to fear of a legal battle with Sikorsky that would make known to the public the degrees of mismanagement and political parrying that have defined the efforts to replace the *Sea King* since cancellation of the NSA project in 1993. As Sharon Hobson has written, the cancellation was "followed by a decade of false starts and promises, political interference and changed rules."[17] In the end, the government actually agreed to give the contractor more money. At the same time that the two-year delay was accepted by the DND and PWGSC in December 2008, the contract was amended to include an extra $117 million for Sikorsky. No information regarding what that money was for was released. The *Sea Kings* will enter their fiftieth year in service in Canada before the fleet of *Cyclones* is delivered. The real reasons for acceptance of the protracted delivery schedule, the refusal to impose damages, and the decision to forward Sikorsky more money will remain speculation.

The *Cyclone* could never have been delivered under the original RFP timeline. The delivery dates are not even close. Nobody knows when the helicopters will be delivered. Sadly, nobody ever has. The original predictions that Sikorsky would be late in delivery did not require profound analysis. The delays were obvious to those in the industry based on what still needed to be done to certify the aircraft. Colonel Akitt concluded his paper with the sad observation that, "based upon the evidence to date and the failure of civil-military relations in this instance, there is little chance of the project providing for the needs of the Canadian Forces."[18] The Liberal government's choice of Sikorsky prevented the timely procurement of an aircraft that was long overdue for the CF.

In his book about his time in office, Jean Chrétien brought up his cancellation of the NSA project. The two paragraphs, however, only continued the spin that he used when he was Prime Minister. He maintained that it was in the interest of saving money. But considering that the helicopters had always been deemed a necessary capability by the CF and that even Chrétien had to admit the *Sea Kings* needed to be replaced, the cancellation really meant that another government would have to pay more later for new helicopters. A procurement that has taken over thirty years and counting has not saved Canada money. As General George Macdonald, former Vice Chief of the Defence Staff and Director General Aerospace Development at the time of the NSA cancellation, explained, "I shudder to think of the total man hours involved in replacing the Sea King – even since 1994."[19] Chrétien went on to use the spurious logic that the *Sea Kings* could not have been as useless as the military claimed them to be because the same model was used in the United States to "ferry the president from the White House to Camp David."[20] Even out of office, Chrétien continued to ignore the actual function of the helicopters. In Canada, their function is not to be a glorified taxi for a head of state. That is not a measure of their capability. They are meant to operate in modern military operations at sea and to monitor and protect Canadian coastal approaches. In their actual roles, the *Sea Kings* had ceased to be effective even before Chrétien took office.

The more recent procurement process, as of 2008, has also incurred a great deal of criticism. It has been quite similar to the MHP. Sculpting of the procurement before the RFP was still essential. As the accusations went, the SOR and RFP were drafted in such a way that only one company could meet the requirements. As a result, there was no need for a competition like that for the MHP. The government could "sole-source" the contract to that specific company unilaterally. The advantage to this approach is that it did not take six years to write the SOR, another four to sign a contract, and another seven to deliver the product, as in the case of the MHP. The procurement timeline was noticeably shorter. The ability of a company to deliver equipment as soon as possible to get the CF the equipment that it needed was vital to that company being selected. But the pros and cons of sole-sourcing are a story for another day.[21]

The lessons learned by defence officials during the NSA/NSH project regarding the use of developmental technology did not result in the expedient procurement of a new maritime helicopter as the government still controlled the process. It was in its power to repeat the mistake. Any one procurement decision by itself did not create a case against the government. But the list of decisions, reversals, and oddities regarding the procurement proves that it was not done in an unbiased and ethical fashion. In Canada, civilians are in control of defence procurement and make decisions based on what they think the population wants. In 1993, they were correct in assuming that Canadians cared little about a major investment in military equipment during a recession, regardless of whether that equipment was deemed necessary by the military. They did not know at the time what the repercussions of cancelling the NSA contract would be. And citizens have a right to be wrong. This is true of governments as well. But they should be held accountable for mismanagement and deception. Most of the officials involved in the MHP, however, will have long since retired or moved on before the project is complete. The combined attempts to replace the *Sea King* helicopters have formed the longest-running military acquisition effort in Canadian history and the true effects of the 1993 NSA cancellation and the political direction of the MHP have yet to be fully realized.

Notes

Preface

Epigraph: General (retired) Paul Manson, interview with Aaron Plamondon, 24 September 2005.

1 Douglas Bland, ed., *Transforming National Defence Administration* (Kingston: School of Policy Studies, Queen's University, 2005), 61.

2 Joseph Jockel, *The Canadian Forces: Hard Choices, Soft Power* (Toronto: Canadian Institute of Strategic Studies, 1999), 74-75.

3 Martin Shadwick, "Replacing the *Sea King*: Canada Examines the Need to Replace Its *Sea Kings* with a New ASW Helicopter," *Wings, Canada's Navy Annual* Commemorative Issue (1985): 164.

4 Marc Milner, *Canada's Navy: The First Century* (Toronto: University of Toronto Press, 1999), 226.

5 This and the rest of the paragraph are based on Dan Middlemiss, "Defence Procurement in Canada," in *Canada's International Security Policy,* ed. David B. Dewitt and David Leyton-Brown (Scarborough: Prentice-Hall, 1995), 401-4.

6 Stephen Martin, ed., *The Economics of Offsets: Defence Procurement and Countertrade* (Amsterdam: Routledge, 1996), 3. Martin explained that by 1996, "over 130 countries had some form of offset policy."

7 Robert R. Reford, "The Public and Public Policy. The Impact of Society on the Canadian Policy Process," in *Canada's International Security Policy,* ed. Dewitt and Leyton-Brown, 313. On nuclear weapons in Canada, see Andrew Richter, *Avoiding Armageddon: Canadian Military Strategy and Nuclear Weapons, 1950-63* (Vancouver: UBC Press, 2002); Brian Buckley, *Canada's Early Nuclear Policy: Fate, Chance, and Character* (Montreal and Kingston: McGill-Queen's University Press, 2000); and Erika Simpson, *NATO and the Bomb: Canadian Defenders Confront Critics* (Montreal and Kingston: McGill-Queen's University Press, 2001).

8 PWGSC, "Settlement Reached with EH Industries for the EH-101 Helicopter Program," news release, 23 January 1996.

9 David Bercuson, "Time to Wake Up on Procurement," *Legion Magazine* (November 2005): retrieved from www.legionmagazine.com.

10 Colonel (retired) John Cody, former CO of 423 Helicopter Squadron, e-mail to Aaron Plamondon, 5 March 2006.

11 Colonel (retired) Laurence McWha, former CO of 423 Helicopter Squadron, e-mail to Aaron Plamondon, 10 March 2006.

12 General (retired) Paul Manson, "Procurement Cycle Growth: The Race between Obsolescence and Acquisition of Military Equipment in Canada, 1960 to the Present," paper presented at the Canadian Institute of Strategic Studies Seminar, 22 July 2005, http://www.cda-cdai.ca/.

13 Colonel (retired) John Cody, former CO of 423 Helicopter Squadron, e-mail to Aaron Plamondon, 12 January 2006.

14 Jim Brown, "Full Circle: Electoral Promise to Axe Choppers Returns to Haunt Chrétien," Canadian Press Newswire, 5 January 1998.
15 Colonel Brian Akitt, "The *Sea King* Replacement Project: A Lesson in Failed Civil-Military Relations," Canadian Forces College, 2003, 23; paper obtained from Colonel Akitt.
16 "Janet Breen, on her own behalf and on behalf of the estate of Peter Breen, et al. vs. Keystone Helicopter Corporation, et al.," Court of Common Pleas, Philadelphia County, Case ID: 090601841. Retrieved from http://beta.images.theglobeandmail.com/archive/00079/Sikorsky_suit_79387a.pdf.

Introduction

1 Peter Diekmeyer, "Defence Procurement in Canada: Is the System Broken?" *Canadian Defence Review* 14, 6 (2008): 8. For more on the procurement system in Canada, see Aaron Plamondon, *Equipment Procurement in Canada and the Civil-Military Relationship: Past and Present* (Calgary: Centre for Military and Strategic Studies, 2008); and Alan Williams, *Reinventing Canadian Defence Procurement: A View from the Inside* (Montreal and Kingston: McGill-Queen's University Press, 2006).
2 For more on this subject see A. Crosby, "Project Management in DND," *Canadian Defence Quarterly* 18, 6 Special No. 2 (1989): 60.
3 Douglas Bland, ed., *Transforming National Defence Administration* (Kingston: School of Policy Studies, Queen's University, 2005), 61.
4 For a more thorough discussion of the nature of Canadian defence procurement during the first half of the twentieth century, see Aaron Plamondon, "Casting Off the Yoke: The Transition of Canadian Defence Procurement within the North Atlantic Triangle, 1907-1953," MA thesis, Royal Military College of Canada, 2001.
5 See "Procurement Service Concepts," a lecture presented by R.M. Brophy, Deputy Minister, Department of Defence Production, to the Industrial College of the Armed Forces, Washington, DC, 12 December 1952, http://www.ndu.edu/library/ic2/L53-071.pdf.
6 Memorandum by Deputy Minister L.R. LaFleche, 20 January 1938, Library and Archives Canada (LAC), RG24, reel C-5091.
7 C.P. Stacey, *Military Problems of Canada: A Survey of Defence Policies and Strategic Conditions Past and Present* (Toronto: Canadian Institute of International Affairs; Ryerson Press, 1940), 127.
8 C.P. Stacey, *Arms, Men, and Governments: The War Policies of Canada, 1939-45* (Ottawa: Queen's Printer, 1970), 101-2.
9 Royal Commission on the Bren Machine Gun Contract, *Report of the Royal Commission on the Bren Machine Gun Contract,* 29 December 1938, LAC, RG24, reel C-5091.
10 This and the following paragraph were retrieved from Brophy, "Procurement Service Concepts."
11 See "Administrative History," Department of Defence Production Fonds, LAC, RG49M 69/9157, http://www.collectionscanada.gc.ca/.
12 Brophy, "Procurement Service Concepts."
13 See DDP Fonds, LAC, RG49M 69/9157, http://www.collectionscanada.gc.ca/.
14 Ibid.
15 Brophy, "Procurement Service Concepts."
16 This and the rest of the paragraph were retrieved from DDP Fonds, LAC, RG49M 69/9157, http://www.collectionscanada.gc.ca/.
17 See Chapter 4.
18 Pierre L. Lagueux, ADM Mat, DND, presentation to SCONDVA, 2 March 1999, http://www.parl.gc.ca/.
19 See PWGSC web page, http://www.tpsgc-pwgsc.gc.ca/.
20 Crosby, "Project Management in DND," 60.
21 Unless otherwise referenced, this account of the process described after 1993 was retrieved from Lagueux, presentation to SCONDVA.
22 For more on this issue, see Chapter 2 of this volume.
23 Demetrios Xenos, "Canada's Industrial and Regional Benefits Policy" (Ottawa: Industry Canada, 2005), retrieved from the ADM Mat website at http://74.125.155.132/search?

q=cache:8WuuY6uRewsJ:www.forces.gc.ca/admmat/hcmfelex-mchpdvdf/documents/Canadas_Industrial_and_Regional_Benefits_Policy.pdf+Canada%E2%80%99s+Industrial+and+Regional+Benefits+Policy+xenos&cd=1&hl=en&ct=clnk.

24 Martin, *The Economics of Offsets*, 2.

25 Xenos, "Canada's Industrial and Regional Benefits Policy,"ADM (Materiel) official website at http://www.forces.gc.ca/admmat/hcmfelex-mchpdvdf/documents/Canadas_Industrial_and_Regional_Benefits_Policy.pdf.

26 CASR, "LOIs – Letters of Interest," http://www.casr.ca/.

Chapter 1: Procurement in Canada

Epigraphs: Desmond Morton, *Understanding Canadian Defence* (Toronto: Penguin Canada, 2003), 82; C.P. Stacey, *The Military Problems of Canada: A Survey of Defence Policies and Strategic Conditions Past and Present* (Toronto: Canadian Institute of International Affairs; Ryerson Press, 1940), 125-26.

1 See Aaron Plamondon, "Casting Off the Yoke: The Transition of Canadian Defence Procurement within the North Atlantic Triangle, 1907-53" (MA thesis, Royal Military College of Canada, 2001).

2 David Bercuson, "Time to Wake Up on Procurement," *Legion Magazine* (November 2005), retrieved from www.legionmagazine.com.

3 See Desmond Morton, *Ministers and Generals: Politics in the Canadian Militia, 1868-1904* (Toronto: University of Toronto Press, 1970); Stephen Harris, *Canadian Brass: The Making of a Professional Army, 1860-1939* (Toronto: University of Toronto Press, 1988); Richard Preston, *Canada and "Imperial Defense": A Study of the Origins of the British Commonwealth's Defense Organisation, 1867-1919* (Durham: Duke University Press, 1967); Donald C. Gordon, *The Dominion Partnership in Imperial Defence, 1870-1914* (Baltimore: Johns Hopkins University Press, 1965); Floyd Low, "Canada Militia Policy, 1885-1914," *Canadian Defence Quarterly* 11, 2 (Autumn 1981): 29-38; and John G. Armstrong, "The Dundonald Affair," *Canadian Defence Quarterly* 11, 2 (Autumn 1981): 39-45.

4 Stacey, *The Military Problems of Canada*.

5 See R.G. Haycock, "Early Canadian Weapons Acquisition: 'That Damned Ross Rifle,'" *Canadian Defence Quarterly* 14, 3 (Winter, 1984-85): 48-57; David Carnegie, *The History of Munitions Supply in Canada 1914-1918* (London: Longmans Green, 1925); R.G. Haycock, "Policy, Patronage, and Production: Canada's Public and Private Munitions Industry in Peacetime, 1867-1939," in *Canada's Defence Industrial Base: The Political Economy of Preparedness and Procurement*, ed. David G. Haglund (Kingston: R.P. Frye, 1988); Roger F. Phillips, François J. Dupuis, and John A. Chadwick, *The Ross Rifle Story* (Sydney, NS: John A. Chadwick, 1984); A. Fortescue Duguid, *A Question of Confidence: The Ross Rifle in the Trenches* (Ottawa: Service Publications, 2000); and Bill Rawling, *Surviving Trench Warfare: Technology and the Canadian Corps, 1914-1918* (Toronto: University of Toronto Press, 1992).

6 J.L. Granatstein, *Arming the Nation: Canada's Industrial War Effort, 1939-45* (Ottawa: Council of Chief Executives, 2005); see also Michael Hennessy, "The Industrial Front: The Scale and Scope of Canadian Industrial Mobilization during the Second World War," in *Forging a Nation: Perspectives on the Canadian Military Experience*, ed. B. Horn (St. Catharines: Vanwell, 2002). On the Canadian-American Defence Production and Development Sharing Agreements, see John J. Kirton, "The Consequences of Integration: The Case of the Defence Production Sharing Agreements," in *Continental Community? Independence and Integration in North America*, ed. W. Andrew Axline (Toronto: McClelland and Stewart, 1974); Lawrence Aronsen, "Canada's Post-War Re-Armament: Another Look at American Theories of the Military Industrial Complex," *Canadian Historical Association Historical Papers* (1981); and D.W. Middlemiss, "A Pattern of Cooperation: The Case of the Canadian American Defence Production and Development Sharing Arrangements, 1958-1963" (PhD diss., University of Toronto, 1975). Other sources that deal with post-Second World War Canadian defence economics are William J. Yost, *Industrial Mobilization in Canada* (Ottawa: Conference of Defence Associations, 1983); D.W. Middlemiss, "Canada and Defence Industrial Preparedness: A Return to Basics," in *Canada's Defence: Perspectives on Policy in the Twentieth Century*, ed. B.D. Hunt and R.G. Haycock (Toronto: Copp Clark Pitman, 1993); John Treddenick,

The Economic Significance of the Canadian Defence Industrial Base, Centre for Studies in Defence Resource Management Report 15 (Kingston: Royal Military College of Canada, 1987); Lawrence R. Aronsen, "From World War to Limited War: Canadian-American Industrial Mobilization for Defence," *Revue internationale d'histoire militaire* 51 (1982); Robert Van Steenburg, "An Analysis of Canadian-American Defence Economic Cooperation: The History and Current Issues," in *Canada's Defence Industrial Base: The Political Economy of Preparedness and Procurement,* ed. David G. Haglund (Kingston: R.P. Frye, 1988); and Joseph Jockel, *No Boundaries Upstairs: Canada, the United States, and the Origins of North American Air Defence, 1945-1958* (Vancouver: UBC Press, 1987). Among the many books on the *Arrow,* see Greig Stewart, *Shutting Down the National Dream: A.V. Roe and the Tragedy of the Avro* Arrow (Toronto: McGraw-Hill Ryerson, 1988).

7 Dan Middlemiss, "Defence Procurement in Canada," in *Canada's International Security Policy,* ed. David B. Dewitt and David Leyton-Brown (Scarborough: Prentice Hall, 1995), 391.

8 William Johnston, "Canadian Defence Industrial Policy and Practice: A History," *Canadian Defence Quarterly* 18, 6 Special no. 2 (June 1989): 21.

9 Donald Creighton, *Macdonald: The Young Politician – the Old Chieftain* (Toronto: University of Toronto Press, 1998), 325-32.

10 See Richard A. Preston, *The Defence of the Undefended Border: Planning for War in North America, 1867-1939* (Montreal and Kingston: McGill-Queen's University Press, 1977), 26.

11 John Hasek, *The Disarming of Canada* (Toronto: Key Porter, 1987), 105.

12 Haycock, "Early Canadian Weapons Acquisition," 89.

13 Canada, House of Commons, *Debates,* 1905, 9122, and 1908, 8980-94. On the history of the Ross rifle, see Phillips, Dupuis, and Chadwick, *The Ross Rifle Story*.

14 Haycock, "Policy, Patronage, and Production," 72.

15 Haycock, "Early Canadian Weapons Acquisition," 48-51.

16 This quote and the rest of the paragraph were retrieved from "Extracts from Minutes of the Imperial Conference on the Naval and Military Defence of the Empire, 1909," *Documents on Canadian External Relations* (*DCER*), vol. 1, 1909-18 (Ottawa: Department of External Relations, 1967), 231-32.

17 See Phillips, Dupuis, and Chadwick, *The Ross Rifle Story*; and Haycock, "Early Canadian Weapons Acquisition."

18 George Stanley, *Canada's Soldiers: The Military History of an Unmilitary People* (Toronto: Macmillan, 1974), 317.

19 Haycock, "Early Canadian Weapons Acquisition," 55.

20 Desmond Morton and J.L. Granatstein, *Marching to Armageddon: Canada and the Great War 1914-1919* (Toronto: Lester and Orpen Dennys, 1989), 81; see also Canadian Liberal Party, *War Contract Scandals, as Investigated by the Public Accounts Committee of the House of Commons, 1915; also the Purchase of Boots, as Investigated by the Special "Boot Committee" Appointed by the House of Commons* (Ottawa: privately printed, 1915).

21 J.A. Foster, *Muskets to Missiles: A History of Canada's Ground Forces* (Toronto: Methuen, 1987), 100.

22 Desmond Morton, *When Your Number's Up: The Canadian Soldier in the First World War* (Toronto: Random House, 1994), 33.

23 This was the conclusion of the British commissioners who investigated the Canadian munitions industry. See Carnegie, *The History of Munitions Supply,* 2-8.

24 Minutes of the Imperial War Conference of 1917, 26 March 1917, LAC, RG25, vol. 2279, Conference file, 31.

25 For more on this, see Aaron Plamondon, "The Great War and Imperial Weapons and Equipment Standardization," in *Perspectives on War: New Views on Historical and Contemporary Security Issues,* ed. J.P. Marchant and Chris Bullock (Calgary: Centre for Military and Strategic Studies, 2003).

26 Air Board Committee Minutes, Meeting No. 216, 9 January 1922, LAC, RG24, vol. 3192.

27 Air Board Committee Minutes, Meeting No. 261, 12 May 1922, LAC, RG24, vol. 3192.

28 Imperial Conference of 1926, Summary of Proceedings, Christie Papers, LAC, MG30 E15, vol. 12, reel C-3883.

29 For more on the Bren gun scandal, see the Introduction to this volume.

30 For more on the Defence Purchasing Board, see Canada, House of Commons, Bill 38, *Debates,* vol. 4, 1939, LAC, RG24, reel C-5091.
31 C.P. Stacey, *Arms, Men, and Governments: The War Policies of Canada, 1939-45* (Ottawa: Queen's Printer, 1970), 102; for the joint ownership of the plant by the Canadian and British governments, see Memorandum from L.R. LaFleche, Deputy Minister, to Colonel D.E. Dewar, Master General of the Ordnance, 18 January 1938, LAC, RG24, reel C-5091.
32 Randall Wakelam, "Flights of Fancy: RCAF Fighter Procurement, 1945-54" (MA thesis, Royal Military College of Canada, 1997), 2.
33 H. Duncan Hall and C.C. Wrigley, *Studies of Overseas Supply* (London: H.M. Stationery, 1956), 91-107.
34 Stacey, *Arms, Men, and Governments,* 487.
35 Granatstein, *Arming the Nation*; see also Hennessy, "The Industrial Front."
36 Gilbert Tucker, *The Naval Service of Canada, II* (Ottawa: King's Printer, 1952), 60; Stacey, *Arms, Men, and Governments,* 487.
37 W.G.D. Lund, "The Royal Canadian Navy's Quest for Autonomy in the North West Atlantic," in *RCN in Retrospect, 1910-68,* ed. James A. Boutilier (Vancouver: UBC Press, 1982), 156.
38 This and the rest of the paragraph were retrieved from David Zimmerman, *The Great Naval Battle of Ottawa: How Admirals, Scientists, and Politicians Impeded the Development of High Technology in Canada's Wartime Navy* (Toronto: University of Toronto Press, 1989), 3-8. Quotations are on page 5.
39 Canadian Forces, "The Armoured Corps Story," *Sentinel* 2, 5 (1966): 32-33.
40 Granatstein, *Arming the Nation,* 11-12; for more on the history of the *Ram* tank, see Clive M. Law, *Making Tracks: Tank Production in Canada* (Ottawa: Service Publications, 2001); and Graham Broad, "'Not Competent to Produce Tanks': The Ram and Tank Production in Canada, 1939-1945," *Canadian Military History* 11 (2002).
41 David J. Bercuson, *True Patriot: The Life of Brooke Claxton, 1898-1960* (Toronto: University of Toronto Press, 1993), 179.
42 For a full discussion on the Canadian-American defence relationship during the war, see S.D. Pierce and A.F.W. Plumptre, "Canada's Relations with War Time Agencies in Washington," *Canadian Journal of Economics and Political Science* 11, 3 (1945); R.D. Cuff and J.L. Granatstein, *Canadian-American Relations in Wartime: From the Great War to the Cold War* (Toronto: Hakkert, 1975); and H. Duncan Hall, *North American Supply* (London: H.M. Stationery Office; Longmans Green, 1955).
43 J.L. Granatstein, "The American Influence on the Canadian Military, 1939-1963," *Canadian Military History* 2, 1 (1993): 66.
44 It was first proposed at a meeting of the Permanent Joint Board of Defence in 1945. See Stanley Dzuiban, *Military Relations between the United States and Canada, 1939-1945,* United States Army in World War II Special Studies (Washington, DC: US Government Printing Office, 1959), 334-35; for the actual agreement, see "Joint Statement by the Governments of Canada and the United States of America Regarding Defence Co-Operation between the Two Countries and Statement by the Prime Minister Mr. W.L. Mackenzie King," 12 February 1947, in *Canadian Foreign Policy 1945-54: Selected Speeches and Documents,* ed. R.A. MacKay (Toronto: McClelland and Stewart, 1970), 228.
45 See Peter Michael Archambault, "The Informal Alliance: Anglo-Canadian Defence Relations, 1945-60" (PhD diss., University of Calgary, 1997).
46 Middlemiss, "A Pattern of Cooperation," 83-85.
47 F.T. Smye, AGM Sales and Contracts A.V. Roe, to G/C J.A. Easton, 27 January 1947, LAC, RG24, vol. 5404, file 60-5-19.
48 Wakelam, "Flights of Fancy," 71.
49 Stewart, *Shutting Down the National Dream,* 144, 120.
50 Wakelam, "Flights of Fancy," 157.
51 Robert Bothwell, "Defense and Industry in Canada 1935-1970," in *War, Business, and World Military-Industrial Complexes,* ed. Benjamin Franklin Cooling (Port Washington, NY: Kennikat Press, 1981), 115.
52 Wakelam, "Flights of Fancy," 201-3.
53 Ibid., 236.

54 C.F.W. Pound, *The Defence Program and National Industrial Development,* Defence Research Analysis Establishment Report 34 (Ottawa: Department of National Defence, 1973), 29.
55 Marc Milner, "A Canadian Perspective on Canadian and American Relations since 1945," in *Fifty Years of Canada-United States Defence Cooperation: The Road from Ogdensburg,* ed. Joel J. Sokolsky and Joseph T. Jockel (Lewiston: Edwin Mellen Press, 1992), 152.
56 See J.H.W. Knox, "An Engineer's Outline of RCN History," in *RCN in Retrospect, 1910-1968,* ed. James A. Boutilier (Vancouver: UBC Press, 1982), 317.
57 Rear Admiral S. Mathwin Davis, "Naval Procurement, 1950-1965," in *Canada's Defence Industrial Base: The Political Economy of Preparedness and Procurement,* ed. David G. Haglund (Kingston: R.P. Frye, 1988), 97.
58 Knox, "An Engineer's Outline of RCN History," 320.
59 Marc Milner, *Canada's Navy: The First Century* (Toronto: University of Toronto Press, 1999), 214-15.
60 This and the rest of the paragraph are based on Michael Hennessy, "The RCN and the Post War Naval Revolution, 1955-1964," paper presented to the Canadian Historical Association Annual Conference, Charlottetown, 1992, 8, 16. See also Davis, "The 'St. Laurent Decision,'" 187-208.
61 Milner, *Canada's Navy,* 222-23.
62 J.L. Granatstein, *Canada's Army: Waging War and Keeping the Peace* (Toronto: University of Toronto Press, 2002), 320.
63 David J. Bercuson, *Blood on the Hills: The Canadian Army in the Korean War* (Toronto: University of Toronto Press, 1999), 23.
64 Ibid., 56, 71; J.L. Granatstein, "The American Influence," 69.
65 Middlemiss, "A Pattern of Cooperation," 183.
66 Stewart, *Shutting Down the National Dream*; James Dow, *The* Arrow (Toronto: James Lorimer, 1997); Palmiro Campagna, *Storms of Controversy: The Secret Avro* Arrow *Files Revealed* (Toronto: Stoddart, 1992); Julius Lukasiewicz, "Canada's Encounter with High-Speed Aeronautics," *Technology and Culture* 27, 2 (1986).
67 Desmond Morton, *Canadian Military History: From Champlain to the Gulf War* (Toronto: McClelland and Stewart, 1992), 240.
68 Middlemiss, "A Pattern of Cooperation," 188.
69 Lieutenant General Guy Simonds, "Where We've Gone Wrong on Defence," *Maclean's,* 23 June 1956: 23, 26.
70 George R. Pearkes, interview with Reginald Roy, 5 April 1967, Special Collections, University of Victoria Library, ACC 74-1, box 5, interview 61.
71 Morton, *Canadian Military History,* 243.
72 For more on the *Heller,* see Peter Michael Archambault, "The Informal Alliance: Anglo-Canadian Defence Relations, 1945-60" (PhD diss., University of Calgary, 1997), 311-18.
73 For more on the *Bobcat,* see ibid., 311-18. Quotation is on page 318.
74 For the complete history of the arrangements, see Middlemiss, "A Pattern of Cooperation," 215; for those who believe that the defence production sharing agreements came about only after Canada cancelled the *Arrow* program and effectively abandoned future domestic development prospects for major weapon systems, see Alistair D. Edgar and David G. Haglund, *The Canadian Defence Industry in the New Global Environment* (Montreal and Kingston: McGill-Queen's University Press, 1995), 62. See also Michael Slack and John Skynner, "Defence Production and the Defence Industrial Base," in *Canada's International Security Policy,* ed. Dewitt and Leyton-Brown, 369-70.

Chapter 2: Early Helicopter Operations

1 Stuart Soward, "Canadian Naval Aviation, 1915-69," in *The RCN in Retrospect, 1910-1968,* ed. James A. Boutilier (Vancouver: UBC Press, 1982), 271-72.
2 Peter Michael Archambault, "The Informal Alliance: Anglo-Canadian Defence Relations, 1945-60" (PhD diss., University of Calgary, 1997), 46.
3 This and the following paragraph are based on Leo Pettipas, *Canadian Naval Aviation, 1945-68* (Winnipeg: Pettipas, 1990), 1-8, 165. For a complete history of the beginnings of Canadian

naval aviation, see Michael Shawn Cafferky, "Towards the Balanced Fleet: A History of the Royal Canadian Naval Air Service, 1943-45" (MA thesis, University of Victoria, 1989).

4 Archambault, "The Informal Alliance," 48.

5 Department of National Defence, *Canada's Defence 1947* (Ottawa: King's Printer, 1947); for more on King, see J.W. Pickersgill and D.F. Forster, *The Mackenzie King Record,* vol. 4 (Toronto: University of Toronto Press, 1960), 6.

6 W.G.D. Lund, "The Rise and Fall of the Royal Canadian Navy 1945-64: A Critical Study of the Senior Leadership, Policy, and Manpower Management" (PhD diss., University of Victoria, 1999), 71.

7 See Pettipas, *Canadian Naval Aviation,* 13-31; and Soward, "Canadian Naval Aviation," 276-77.

8 For a reproduction of this speech in October 1948 and more on Canada's role in the formation of NATO, see Escott Reid, "Canada and the Creation of the North Atlantic Alliance, 1948-1949," in *Canadian Foreign Policy: Historical Readings,* ed. J.L. Granatstein (Toronto: Copp Clark Pitman, 1986), 195.

9 For discussion on what the American and British navies were doing, see Norman Friedman, *The Post War Naval Revolution* (Annapolis: Naval Institute Press, 1986), 9-10, 29.

10 For an analysis of the Canadian response to the postwar naval revolution, see Michael Hennessy, "The RCN and the Post War Naval Revolution, 1955-1964," paper presented to the Canadian Historical Association Annual Conference, Charlottetown, 1992.

11 Minutes of the 387th Meeting of the Naval Staff, 6 October 1947, Directorate of History and Heritage (DHH), Department of National Defence, 1000-100/3.

12 For more on this, see the chapter entitled "Canada and Submarine Warfare, 1909-1950," in Roger Sarty, *The Maritime Defence of Canada* (Toronto: Canadian Institute of Strategic Studies, 1996), 208.

13 Lund, "The Rise and Fall of the Royal Canadian Navy," 119.

14 See Joel Sokolsky, "Canada and the Cold War at Sea, 1945-68," in *RCN in Transition, 1910-1985,* ed. W.A.B. Douglas (Vancouver: UBC Press, 1988), 213; see also Dan Middlemiss, "Economic Considerations in the Development of the Canadian Navy since 1945," in *RCN in Transition, 1910-1985,* ed. W.A.B. Douglas (Vancouver: UBC Press, 1988), 254-79.

15 Brian Cuthbertson, *Canadian Military Independence in the Age of the Superpowers* (Toronto: Fitzhenry and Whiteside, 1977), 127.

16 There was also a personnel shortage to undertake this commitment. See Lund, "The Rise and Fall of the Royal Canadian Navy," ii.

17 Department of National Defence, *Department of Defence Annual Report* for the fiscal year ending March 1949 (Ottawa: King's Printer, 1951), sec. 205, 39.

18 For more on Soviet submarine expansion, see Jan Breemer, *Soviet Submarines: Design, Development, and Tactics* (Surrey, UK: Jane's Information Group, 1989).

19 See Pettipas, *Canadian Naval Aviation,* and Soward, "Canadian Naval Aviation."

20 Soward, "Canadian Naval Aviation," 278; for the quotation, see Peter Charlton, *Nobody Told Us It Couldn't Be Done: The Story of the VX10* (Ottawa: Peter Charlton, 1995), 123.

21 John Chartres, *Westland* Sea King (London: Ian Allan, 1984), 7.

22 "A Chronological Review of Helicopter Operation in the RCN," DHH, 86/377, 1.

23 Minutes of the 399th Meeting of the Naval Staff, 27 January 1948, DHH, 1000-100/3.

24 See the primary author on naval helicopters in Canada, Michael Shawn Cafferky, "Uncharted Waters: The Development of the Helicopter Carrying Destroyer in the Post-War Royal Canadian Navy, 1943-64" (PhD diss., Carleton University, 1996), 116. This dissertation was later published as *Uncharted Waters: A History of the Canadian Helicopter-Carrying Destroyer* (Halifax: Centre for Foreign Policy Studies, Dalhousie University, 2005).

25 "H-5 Helicopter," DHH, 86/377, 45.

26 This was retrieved from Cafferky, "Uncharted Waters," 124, 193.

27 "Staff Requirements – RCN Helicopters," Appendix A to Minutes of the 492nd Meeting of the Naval Staff, 31 July 1951, DHH, 1000-100/3, 1-2.

28 "RCN Helicopters – Phased Introduction Programme," Appendix B to Minutes of the 492nd Meeting of the Naval Staff, 31 July 1951, DHH, 1000-100/3, 1-2.

29 J.A. Foster, *Sea Wings: A History of Waterborne Defence Aircraft* (Toronto: Methuen, 1986), 129.
30 Minutes of the 410th Meeting of the Naval Staff, 29 April 1948, DHH, 1000-100/3.
31 Pettipas, *Canadian Naval Aviation,* 66.
32 David J. Bercuson, "Time to Wake Up on Procurement," *Legion Magazine* (November 2005), retrieved from www.legionmagazine.com.
33 Foster, *Sea Wings,* 129; for a detailed operational history of helicopters used in the role of forest fire suppression, see Glenn Cook, *Vignettes of a Canadian Naval Aviator, 1955-1983: A Memoir* (Ottawa: self-published, 2005), 108-12, 119-47.
34 Pettipas, *Canadian Naval Aviation,* 66-67.
35 Cafferky, "Uncharted Waters," 146.
36 Foster, *Sea Wings,* 129.
37 Hennessy, "The RCN and the Post War Naval Revolution," 6.
38 "Role of Helicopter in A/S Operations and Seaward Defence," Minutes of the 392nd Meeting of the Naval Board (Special), 16 November 1953, DHH, 1000-100/2.
39 Ibid.
40 "Formation of Helicopter Anti-Submarine Squadron," Chiefs of Staff Minutes 549-2, 17 November 1953, DHH, 86/377.
41 Minutes of the 564th Meeting of the Naval Staff, 24 September-9 October 1953, DHH, 1000-100/3.
42 "Formation of Helicopter Anti-Submarine Squadron," Chiefs of Staff Minutes 549-2, 17 November 1953, DHH, 86/377.
43 This and the rest of the paragraph were retrieved from "Role of Helicopter in A/S Operations and Seaward Defence," Minutes of the 392nd Meeting of the Naval Board (Special), 16 November 1953, DHH, 1000-100/2.
44 "A Chronological Review of Helicopter Operation in the RCN," DHH, 86/377, 2.
45 This and the rest of the paragraph were retrieved from "Acquisition of a Second Carrier and Helicopter Ships," Minutes of the 405th Meeting of the Naval Board, 19 May 1954, DHH, 1000-100/2.
46 Minutes of the 476th Meeting of the Naval Board, DHH, 1000-100/2, 3.
47 This and the rest of the paragraph were retrieved from Cafferky, "Uncharted Waters," 183-84, 189-90.
48 This and the rest of the paragraph are based on "ASW Helicopters – Staff Requirements," Minutes of the Meeting of the Naval Staff, 8-24 June 1954, DHH, 1000/100-3, 2-3.
49 For more on how this affected Canada, see "Ad Hoc Committee on the Reappraisal of Current War Plans," Minutes of the Policy Planning Coordinating Committee, 17 September 1956, DHH, 79/249; and "Minutes of Special Meeting," Minutes of the Chiefs of Staff Committee, 26 October 1955, DHH, 73/1223, file 1303.
50 For more on Soviet capabilities, see Breemer, *Soviet Submarines.*
51 Hennessy, "The RCN and the Post War Naval Revolution," 10-12.
52 Cafferky, "Uncharted Waters," 265-66.
53 Hennessy, "The RCN and the Post War Naval Revolution," 12-13.
54 Cafferky, "Uncharted Waters," 240.
55 See Rear Admiral S. Mathwin Davis, "The 'St. Laurent' Decision: Genesis of a Canadian Fleet," in *RCN in Transition, 1910-1985,* ed. W.A.B. Douglas (Vancouver: UBC Press, 1988), 187-208.
56 "A Chronological Review of Helicopter Operation in the RCN," DHH, 86/377, 5-6; on the separate needs of the services, see "Brief on ASW Helicopters in the RCN," [October 1959], DHH, 86/377, 6.
57 "Choice of ASW Helicopter for 1957 Procurement," Minutes of the 476th Meeting of the Naval Board, 8 February 1956, DHH, 1000-100/2, 3.
58 Ibid.
59 This paragraph was retrieved from Cafferky, "Uncharted Waters," 124, 192-93.
60 "Procurement of Helicopters," Minutes of the 112th Meeting of the Policy and Projects Co-Ordinating Committee, 19 August 1957, DHH, 79/246, 2.

61 "Helicopters in Escorts," Meeting of the Naval Secret Staff (NSS) 1115-39, 13 January 1956, LAC, RG24, 83-84/167, vol. 1, file S-8260-11, retrieved from Cafferky, "Uncharted Waters," 91.
62 Peter Charlton and Michael Whitby, *"Certified Serviceable"* Swordfish *to* Sea King*: The Technical Story of Canadian Naval Aviation by Those Who Made It So* (Gloucester, ON: CNATH, 1995), 376-77.
63 Ibid., 61-62.
64 Hennessy, "The RCN and the Post War Naval Revolution," 16-17.
65 Minutes of the 482nd Meeting of the Naval Board, 10 April 1956, DHH, 1000-100/2.
66 Cafferky, "Uncharted Waters," 208-13, 225-30.
67 "Helicopter Summary," 1 February 1957, DHH, 86/377, 1.
68 This and the following paragraph were retrieved from Cafferky, "Uncharted Waters," 220-21; for more on British ASW developments, see Willem Hackman, *Seek and Strike: Sonar, Anti-Submarine Warfare, and the Royal Navy, 1914-54* (London, UK: Science Museum, 1984).
69 This and the rest of the paragraph are based on Charlton, *Nobody Told Us It Couldn't Be Done,* 123-25.
70 Charlton and Whitby, *"Certified Serviceable"* Swordfish *to* Sea King, 377.
71 "Choice of ASW Helicopter for 1957 Procurement," 8 November 1955, LAC, RG24, 84/84/167, vol. 11, file 1115-39, p. 2.
72 Ibid.
73 Cafferky, "Uncharted Waters," 180-81.
74 For more on the recession, see Lawrence H. Officer and Lawrence B. Smith, "Stabilization in the Postwar Period," in *Canadian Economic Problems and Policies,* ed. Lawrence H. Officer and Lawrence B. Smith (Toronto: McGraw-Hill, 1970), 9.
75 Jon B. Mclin, *Canada's Changing Defence Policy, 1957-63: The Problem of a Middle Power in Alliance* (Baltimore: Johns Hopkins University Press, 1967), 120-22.
76 "Brief on ASW Helicopters in the RCN," [October 1959], DHH, 86/377, 2.
77 "The Concept of Employing Anti-Submarine Helicopters from Escort Vessels," Minutes of the 22/57 Meeting of the Naval Staff, 17 September 1957, DHH, 1000-100/3, 2.
78 Cafferky, "Uncharted Waters," 288.
79 Charlton and Whitby, *"Certified Serviceable"* Swordfish *to* Sea King, 377.
80 "Helicopter Trials on HMCS Ottawa," Minutes of the 554th Meeting of the Naval Board, 4 December 1957, DHH, 1000-100/2, 2; for the final report, see "Brief on ASW Helicopters in the RCN" [October 1959] (this event was cited within the document as 4 March 1958), DHH, 86/377, 2.
81 Minutes of the 584th Meeting of the Naval Board, 14-16 January 1959, DHH, 1000-100/2.
82 For more on this, see D.G. Brassington, "The Canadian Development of VDS," *Maritime Warfare Bulletin* Commemorative Edition (1985); and Hennessy, "The RCN and the Post War Naval Revolution," 18.
83 Jerry Proc, "Asdic and Sonar in the RCN," http://jproc.ca/.
84 Marc Milner, *Canada's Navy: The First Century* (Toronto: University of Toronto Press, 1999), 207.
85 For more on the threat, see Operational Research Group, "An Appreciation of the Threat to North America of Submarine-Launched Missiles Carrying Nuclear Weapons," 14 August 1958, LAC, RG24, 83-84/167, vol. 2043, file S-5151-49-1; and J.D.F. Kealy and E.C. Russell, *A History of Canadian Naval Aviation, 1918-62* (Ottawa: Naval Historical Section, DND, 1965), 125.
86 Hennessy, "The RCN and the Post War Naval Revolution," 15.
87 Reginald H. Roy, *For Most Conspicuous Bravery: A Biography of Major-General George R. Pearkes, V.C., through Two World Wars* (Vancouver: UBC Press, 1977), 279.
88 "Progress Report on the RCN Nuclear Submarine Study," Project L-2 NSS 8000-SSN, DHH, 79/246; for a detailed discussion on nuclear submarines in Canada, see Rear Admiral S. Mathwin Davis, "It Has All Happened Before: The RCN, Nuclear Propulsion, and Submarines – 1958-68," *Canadian Defence Quarterly* 17 (1987).

89 "Staff Characteristics for an Escort Borne ASW Helicopter," Minutes of the 176th Meeting of the Policy and Projects Coordinating Committee, 9 September 1959, DHH, 79/246, 4.
90 See Aaron Plamondon, "Casting Off the Yoke: The Transition of Canadian Defence Procurement within the North Atlantic Triangle, 1907-1953" (MA thesis, Royal Military College of Canada, 2001).
91 Cafferky, "Uncharted Waters," 300.
92 Minutes of the 49th Meeting of the Vice Chiefs of Staff Committee, 18 September 1959, DHH, 73/1223, file 1308-1309.
93 "Brief on ASW Helicopters in the RCN," [October 1959], DHH, 86/377, 4.
94 For expected timelines, see "Choice of ASW Helicopter for 1957 Procurement," Minutes of the 476th Meeting of the Naval Board, 8 February 1956, DHH, 1000-100/2, 3; see also "The Concept of Employing Anti-Submarine Helicopters from Escort Vessels," Minutes of the 22/57 Meeting of the Naval Staff, 17 September 1957, DHH, 1000-100/3, 2.

Chapter 3: The Procurement of the *Sea King*

Epigraph: Rear Admiral Jeffry V. Brock, RCN Maritime Commander Atlantic, Address on the Occasion of the Introduction of the *Sea King* Helicopter into the Royal Canadian Navy, 28 August 1963, NSS 8885-15, DHH, 79/246.

1 For the retirement of the Banshee fleet in 1961-62, see Leo Pettipas, *Canadian Naval Aviation, 1945-68* (Winnipeg: Pettipas, 1990), 148.
2 Minutes of the 573rd Meeting of the Naval Board, 17 July 1958, DHH, 1000-100/2; for more on the discussion, see Minutes of the 576th-4 Meeting of the Naval Board, 24 September 1958, DHH, 1000-100/2.
3 "Brief on ASW Helicopters in the RCN," [October 1959], DHH, 86/377, 5.
4 This and the rest of the paragraph are based on the Minutes of the 643rd Meeting of the Naval Board, 27 January 1961, DHH, 1000-100/2, 1. The Naval Staff paper "ASW Helicopter Procurement" was file NS C 7820-102.
5 Minutes of the 643rd Meeting of the Naval Board, 27 January 1961, DHH, 1000-100/2, 2.
6 "Helicopter Summary," 1 February 1957, DHH, 86/377, 9.
7 Jeffry Brock, "Notes on NSS 1279-118," 12th Senior Officers' Meeting, 20 November 1961, DHH, Naval Policy file 1650-1, vol. 3.
8 Michael Hennessy, "The RCN and the Post War Naval Revolution, 1955-1964," paper presented to the Canadian Historical Association Annual Conference, Charlottetown, 1992, 27-28.
9 Ibid., 10-11.
10 Ibid.
11 For the specifics on the 30 percent cost overrun, see "Breakdown of the Cost Increase in the CHSK-1 [the proposed name for the Canadian version]," 18 January 1961, DHH, 79/246, folder 81, B-1.
12 Minutes of the 218th Meeting of the Naval Policy and Coordinating Committee, 15 August 1961, DHH, 79/246, 1.
13 "All Weather Helicopter Operations," n.d., DHH, 86/377, 10-11.
14 This and the rest of the paragraph were retrieved from Minutes of the 643rd Meeting of the Naval Board, 27 January 1961, DHH, 1000-100/2, 2-3.
15 "Advantages and Disadvantages of Helicopters Considered for the DDE Programme," 18 January 1961, DHH, 79/246, folder 81, B-1.
16 Minutes of the 643rd Meeting of the Naval Board, 27 January 1961, DHH, 1000-100/2, 2-3.
17 "RCN Helicopter Procurement Programme," Minutes of the 657th Meeting of the Naval Board, 23 August 1961, DHH, 1000-100/2, 1.
18 Ibid.
19 "ASW Procurement – DDE Programme," Minutes of the 218th Meeting of the Naval Policy Coordinating Committee, 15 August 1961, DHH, 79/246, 2. The Policy and Projects Coordinating Committee of the navy was changed to the Naval Policy Coordinating Committee on 7 June 1960.

20 This and the following sentence are based on "ASW Helicopter Procurement," Minutes of the 217th Meeting of the Naval Policy Coordinating Committee (previously the Policy and Projects Coordinating Committee), 9 August 1961, DHH, 79/246, 1.
21 "ASW Procurement – DDE Programme," Minutes of the 218th Meeting of the Naval Policy Coordinating Committee, 15 August 1961, DHH, 79/246, 2.
22 "Air Operations in Small Ships," 2 May 1962, DHH, 86/381, 1.
23 D.A. Golden, Deputy Minister of the Department of Defence Production, to Secretary of the Treasury Board, 23 February 1962, DHH, 79/246, folder 82, B-2.
24 This and the following quotation are from an analysis of the Department of National Defence within a 1962 report entitled the *Royal Commission on Government Organization: Project 16, Interim Report, Presentation to the Commissioners,* as quoted in Douglas Bland, ed., *Canada's National Defence,* Vol. II: *Defence Organization* (Kingston: School of Policy Studies, 1998), 37.
25 J. MacNeill, Assistant Secretary of the Treasury Board, to D.A. Golden, Deputy Minister of Defence Production, 12 March 1962, DHH, 79/246, folder 82, B-2.
26 Vice Admiral H.S. Rayner, "Procurement of Sikorsky HSS-2 Helicopters," memo to Deputy Minister of National Defence, 7 February 1962, DHH, 79/246, folder 82, B-2.
27 E.B. Armstrong, Deputy Minister of National Defence, to G.G.E. Steele, Secretary of the Treasury Board, 26 February 1962, DHH, 79/246, folder 82, B-2.
28 This and the rest of the paragraph were retrieved from Minister of National Defence, "Procurement of HSS-2 Helicopters," memo to Deputy Minister, n.d., DHH, 79/246, folder 82, B-2.
29 "Landing the CHSS-2 on the DDE," n.d., DHH, 71/278, 3-19.
30 "Statement by the Hon. Douglas Harkness, Minister of National Defence," Directorate of Public Relations, Department of National Defence, 20 November 1962, DHH, 79/246, folder 83 B3, 1-2.
31 Minutes of the Meeting of the Treasury Board, Minute 590367, 26 September 1962, DHH, vol. 3, B3.
32 "Statement by the Hon. Douglas Harkness, Minister of National Defence," Directorate of Public Relations, Department of National Defence, 20 November 1962, DHH, 79/246, folder 83 B3, 1-2.
33 "CHSS-2 'Sea King' Maintenance Training," 1 November 1962, DHH, 79/246, folder 83 B3, 1.
34 D.L. Thompson, Director of the Aircraft Branch, Department of Defence Production, to E.B. Armstrong, Deputy Minister of National Defence, 7 September 1962, DHH, 79/243, folder 83, B3.
35 Douglas Harkness, Minister of National Defence, to George Nowlan, Minister of Finance, 14 September 1962, DHH, 79/243, folder 83, B3.
36 John L. Orr, ed., *"With Eagle Wings," 423: A Canadian Squadron in Peace and War,* 423 Squadron 60th Anniversary History Project (Halifax: Print Atlantic, 2002), 125. United Aircraft of Canada Limited later became Pratt and Whitney.
37 Glenn Cook, *Vignettes of a Canadian Naval Aviator, 1955-1983: A Memoir* (Ottawa: self-published, 2005), 172-73.
38 This and the rest of the paragraph were retrieved from Hennessy, "The RCN and the Post War Naval Revolution," 32-34; his research is based on the report of A.W. Allan, J. Longhurst, and G. Hughes-Adams to J.C. Rutledge, Director of Shipbuilding, DDP, 28 June 1962; Director General of Ships to A.J.C. Pomeroy, Defence Supply Naval Shipbuilding Panel, 14 December 1962; and G.W. Hunter, Deputy Minister of Defence Production, to Secretary of the Treasury Board, 29 January 1963, LAC, RG49 (Defence Production), vol. 1, file 39-N-1521-2.
39 Rear Admiral Jeffry V. Brock, RCN Maritime Commander Atlantic, Address on the Occasion of the Introduction of the *Sea King* Helicopter into the RCN, 28 August 1963, Naval Secret Staff (NSS) 79/246, 8885-15, DHH.
40 This quotation and the rest of the paragraph are based on Brigadier General Colin Curleigh, "The New Maritime Helicopter: Reliability Will Be Crucial," *Canadian Defence Quarterly* 26, 4 (1997): 26-27; see also "Wedding of the *Sea King,*" *Crowsnest* 16, 3-4 (1964).

41 Vice Admiral H.S. Rayner, Chief of the Naval Staff, statement to the Special Committee on Defence, [1963], DHH, 79/246, folder 7.
42 Minutes of the 14th Senior Officers' Conference, Naval Headquarters, Ottawa, 4 February 1964, DHH, 1983-84/167.
43 Curleigh, "The New Maritime Helicopter," 26.

Chapter 4: The *Sea King* in Canada

Epigraph: "Planned Introduction of *Sea King* (CHSS-2) Helicopters to the Fleet," 20 May 1964, NSS 1115-39, DHH, 79/246.

1 This and the following quotations were retrieved from "Introduction of the *Sea King* Helicopters to the Fleet," Minutes of the 263rd Meeting of the Naval Policy Coordinating Committee, 26 February 1963, DHH, 79/246, 2-3.
2 Martin Shadwick, "Replacing the *Sea King,*" *Wings Magazine,* Commemorative Issue (1985): 163.
3 Colonel (retired) John Orr, "Forty Years On: The *Sea King* Saga," *Air Force* (2003): 2; electronic copy given to Aaron Plamondon.
4 John Chartres, *Westland* Sea King (London: Ian Allan, 1984), 19.
5 Directorate of Naval Operational Requirements, "A Staff Study on the Operational Requirement for a Heliporter," 29 March 1963, DHH, 79/246, folder 13.
6 This and the rest of the paragraph are based on "Introduction of the *Sea King* Helicopters to the Fleet," Minutes of the 263rd Meeting of the Naval Policy Coordinating Committee, 26 February 1963, DHH, 79/246, 2.
7 Michael Shawn Cafferky, "Uncharted Waters: The Development of the Helicopter Carrying Destroyer in the Post-War Royal Canadian Navy, 1943-64" (PhD diss., Carleton University, 1996), 329.
8 This and the remainder of the paragraph are based on Minutes of the 14th Senior Officers' Conference, Naval Headquarters, Ottawa, 4-6 February 1964, DHH, D11D3-12, E-14-15, 3.
9 Peter Charlton, *Nobody Told Us It Couldn't Be Done: The Story of the VX10* (Ottawa: Peter Charlton, 1995), 127.
10 "Landing the CHSS-2 on the DDE," n.d., DHH, 71/278, 5.
11 Research Control Committee, "Quick Securing Device for Helicopter," brief for item 130-5, DHH, 79/246, folder 85, B6, 1-2.
12 Peter Charlton and Michael Whitby, *"Certified Serviceable"* Swordfish *to* Sea King*: The Technical Story of Canadian Naval Aviation by Those Who Made It So* (Gloucester, ON: CNATH, 1995), 381.
13 Charlton, *Nobody Told Us It Couldn't Be Done,* 130.
14 Charlton and Whitby, *"Certified Serviceable"* Swordfish *to* Sea King, 383.
15 Research Control Committee, "Quick Securing Device for Helicopter," brief for item 130-5, DHH, 79/246, folder 85, B6, 1-2.
16 This and the rest of the paragraph are based on Minutes of the 14th Senior Officers' Conference, Naval Headquarters, Ottawa, 4-6 February 1964, DHH, D11D3-12, E-14-15, 3.
17 For more on all the technical specifics of the alterations made, such as to the control system, as a result of the trials, see Charlton and Whitby, *"Certified Serviceable"* Swordfish *to* Sea King.
18 Charlton, *Nobody Told Us It Couldn't Be Done,* 119.
19 "Planned Introduction of *Sea King* (CHSS-2) Helicopters to the Fleet," 20 May 1964, NSS 1115-39, DHH, 79/246, folder 30, 2. In 1965, the total number to be procured was changed to forty-one.
20 Discussion with Commander Porter on future helicopter requirements, [1965], DHH, 86/381.
21 This and the rest of the paragraph were retrieved from Lieutenant Colonel Ross Fetterly, "The Influence of the Environment on the 1964 Defence White Paper," *Canadian Military Journal* 5, 4 (2004-5), http://www.journal.forces.gc.ca/.
22 Ad Hoc Committee on Defence Policy, *The Canadian Defence Budget* (Ottawa: Department of National Defence, 1963), 11.
23 Fetterly, "The Influence of the Environment."

24 Paul Hellyer, *Damn the Torpedoes: My Fight to Unify Canada's Armed Forces* (Toronto: Force 10 Publications, 1990), 46.
25 A. Keith Cameron, "The Royal Canadian Navy and the Unification Crisis," in *The RCN in Retrospect, 1910-1968,* ed. James A. Boutilier (Vancouver: UBC Press, 1982), 335.
26 Fetterly, "The Influence of the Environment."
27 Ad Hoc Committee on Defence Policy, *Report of the Ad Hoc Committee on Defence Policy* (Ottawa: Department of National Defence, 1963), 23.
28 Fetterly, "The Influence of the Environment."
29 Cameron, "The Royal Canadian Navy and the Unification Crisis," 339.
30 Charlton and Whitby, *"Certified Serviceable"* Swordfish *to* Sea King, 116.
31 Ibid., 116, 378, 387-88.
32 Cafferky, "Uncharted Waters," 343.
33 Lieutenant Commander R.L. Rogers, presentation to Royal Netherlands Navy, 7 September 1966, DHH, 79/246, folder 85, B6, 1-3.
34 Charlton and Whitby, *"Certified Serviceable"* Swordfish *to* Sea King, 116.
35 Retrieved from the log book of Colonel (retired) Laurence McWha, 14 January 2007.
36 "Beartrap," *Armed Forces and Military Procurement Review* (1968).
37 Charlton, *Nobody Told Us It Couldn't Be Done,* 138.
38 This and the rest of the paragraph were retrieved from Stuart E. Soward, *Hands to Flying Stations: A Recollective History of Canadian Naval Aviation,* vol. 2 (Victoria: Neptune Developments, 1995), 401.
39 As recounted by Colonel (retired) Laurence McWha, who participated in the exercise.
40 John L. Orr, ed., *"With Eagle Wings," 423: A Canadian Squadron in Peace and War,* 423 Squadron 60th Anniversary History Project (Halifax: Print Atlantic, 2002), 125.
41 See Leo Pettipas, *Canadian Naval Aviation, 1945-68* (Winnipeg: Pettipas, 1990), 153-56; and J.A. Foster, *Sea Wings: A History of Waterborne Defence Aircraft* (Toronto: Methuen, 1986), 129.
42 Soward, *Hands to Flying Stations,* 394-95.
43 Orr, "Forty Years On," 3.
44 David Gibbings, Sea King*: 21 Years Service with the Royal Navy, 1969-1990* (Seaborough Hill, UK: Skylark Press, 1990), 5 7.
45 Ibid.
46 Colonel (retired) Lee Myrhaugen, e-mail to Aaron Plamondon, 28 July 2006.
47 Al Adcock, *H-3* Sea King *in Action* (Corrollton, TX: Signal Publications, 1995), 23.
48 Charlton and Whitby, *"Certified Serviceable"* Swordfish *to* Sea King, 117.
49 Soward, *Hands to Flying Stations,* 283-84; Brigadier General Colin Curleigh, "The New Maritime Helicopter: Reliability Will Be Crucial," *Canadian Defence Quarterly* 26, 4 (1997): 27.
50 Harb Harzan, e-mail to Aaron Plamondon, 21 July 2006. Harzan was a lieutenant in the navy in 1970.
51 Charlton and Whitby, *"Certified Serviceable"* Swordfish *to* Sea King, 120.
52 Soward, *Hands to Flying Stations,* 283. He is referring to Canadian Forces Headquarters, as NDHQ was not formed until 1972.
53 J. Allan Snowie, *The Bonnie: HMCS* Bonaventure (Erin, On: Boston Mills Press, 1987), 261.
54 For a discussion on how there were not enough trained personnel to operate the *Bonaventure* from the beginning and how this harmed RCN capabilities by taking up too much of the naval budget, see W.G.D. Lund, "The Rise and Fall of the Royal Canadian Navy 1945-64: A Critical Study of the Senior Leadership, Policy, and Manpower Management" (PhD diss., University of Victoria, 1999), 391, 435-36, 532.
55 Ibid., 282-83.
56 Colonel (retired) John Orr, e-mail to Aaron Plamondon, 12 December 2005.
57 Michael Hennessy, "Fleet Replacement and the Crisis of Identity," in *A Nation's Navy: In Quest of Canadian Naval Identity,* ed. Rob Huebert, Michael L. Hadley, and Fred W. Crickard (Montreal and Kingston: McGill-Queen's University Press, 1996), 133.
58 The Chief of the Naval Staff had stated that the use of helicopters could be "a marked step toward making up the deficiencies in the lack of surface escorts." See "Formation of Helicopter Anti-Submarine Squadron," Chiefs of Staff Minutes 549-2, 17 November 1953, DHH, 86/377.

59 Soward, *Hands to Flying Stations,* 411.
60 Orr, "Forty Years On," 3.
61 Soward, *Hands to Flying Stations,* 394-95.
62 Charlton, *Nobody Told Us It Couldn't Be Done,* 152.
63 Soward, *Hands to Flying Stations,* 445.
64 Orr, "Forty Years On."
65 Charlton and Whitby, *"Certified Serviceable"* Swordfish *to* Sea King, 401; for more anecdotes on the level of acceptance of unification, see 401-10.
66 Gordon Moyer, phone interview by Aaron Plamondon, 20 June 2004.
67 This and the rest of the paragraph are based on *The Management of Defence in Canada,* 7 July 1972. The full text is in Douglas Bland, ed., *Canada's National Defence,* Vol. II: *Defence Organization* (Kingston: School of Policy Studies, 1998). The quotation is on page 228-29. See also Douglas Bland, *The Administration of Defence Policy in Canada, 1947-85* (Kingston: Ronald P. Frye, 1987), 79-81, 136.
68 Colonel (retired) John Orr, e-mail to Aaron Plamondon, 12 December 2005.
69 Bland, *Canada's National Defence,* 59, 285-86.
70 Charlton and Whitby, *"Certified Serviceable"* Swordfish *to* Sea King, 395.
71 This quotation and the previous four sentences were retrieved from Marc Milner, *Canada's Navy: The First Century* (Toronto: University of Toronto Press, 1999), 265-66. The quotation is on page 266.
72 For more on this, see Gerald Porter, *In Retreat: The Canadian Forces in the Trudeau Years* (Ottawa: Deneau and Greenberg, 1978), 75-78. Porter includes many quotations from varied sources about the ships.
73 Orr, "Forty Years On," 4.
74 *Task Force on Review of Unification of the Canadian Forces: Final Report 15 March 1980,* quoted in Bland, *Canada's National Defence,* 297.
75 This and the following paragraph were retrieved from the official "History of the 423 Squadron," http://www.shearwateraviationmuseum.ns.ca/. The source material for the 423 Squadron history comes from annual historical reports on file at DHH in Ottawa and the 423 Squadron Library in Shearwater.
76 Snowie, *The Bonnie,* 217.
77 *Task Force on Review of Unification of the Canadian Forces: Final Report 15 March 1980,* quoted in Bland, *Canada's National Defence,* 309.
78 This occurred in September 1975. Orr, "Forty Years On," 4.
79 Colonel (retired) Laurence McWha, e-mail to Aaron Plamondon, 4 December 2005.
80 This and the following paragraph were retrieved from "History of 423 Helicopter Squadron," www.shearwateraviationmuseum.ns.ca/squadrons/.

Chapter 5: The New Shipborne Aircraft Project

Epigraph: Department of National Defence, *White Paper on Defence, 1987* (Ottawa: Queen's Printer, 1987), 51.

1 Colonel (retired) John Orr, "Forty Years On: The *Sea King* Saga," *Air Force* (2003): 5; electronic copy given to Aaron Plamondon.
2 Martin Shadwick, "Replacing the *Sea King:* Canada Examines the Need to Replace Its *Sea Kings* with a New ASW Helicopter," *Canada's Navy* (1985): 164-65; John L. Orr, ed., *"With Eagle Wings," 423: A Canadian Squadron in Peace and War,* 423 Squadron 60th Anniversary History Project (Halifax: Print Atlantic, 2002), 125.
3 David Underwood, "The Eyes and Ears of NSA," *Canadian Aviation* (1987): 22.
4 Colonel (retired) John Cody, e-mail to Aaron Plamondon, 20 December 2005.
5 Thomas Lynch, "Naval Shipborne Aircraft: Rotary Flight after the *Sea King,*" *Wings* (1986): 98.
6 "Briefing Note – the Maritime Helicopter Project," 3 May 2001, obtained by Colonel (retired) Laurence McWha.
7 Stephen Priestley, "Politics, Procurement Practices, and Procrastination: The Quarter-Century *Sea King* Helicopter Replacement Saga," *Canadian-American Strategic Review,* retrieved from http://www.casr.ca/ft-mhp1.htm.

8 This and the following paragraph were retrieved from Dan Middlemiss, "Defence Procurement in Canada," in *Canada's International Security Policy,* ed. David B. Dewitt and David Leyton-Brown (Scarborough: Prentice Hall, 1995), 401-2.
9 Marc Milner, *Canada's Navy: The First Century* (Toronto: University of Toronto Press, 1999), 277.
10 As quoted in *The Nova Scotian,* 4-10 February 1978, reprinted in the *Halifax Sunday Herald,* 10 February 2008.
11 Orr, "Forty Years On," 6.
12 Chronology prepared by Colonel (retired) Laurence McWha, 7 December 2000.
13 Gordon Moyer, telephone interview by Aaron Plamondon, 12 August 2004.
14 Gordon Davis, *The Maritime Helicopter Project: The Requirement for a Capable Multi-Purpose* Sea King *Replacement* (Halifax: Centre for Foreign Policy Studies, 2000), 14.
15 David Gibbings, Sea King*: 21 Years Service with the Royal Navy, 1969-1990* (Sea Borough Hill: Skylark Press, 1990), 10-14.
16 Al Adcock, *H-3* Sea King *in Action* (Corrollton, TX: Signal Publications, 1995), 11, 19, 31, 39, 46.
17 Colonel (retired) Laurence McWha, e-mail to Aaron Plamondon, 14 July 2006.
18 Lieutenant Colonel (retired) John W. McDermott in a speech written to 423 Squadron for their 60th anniversary, n.d., copy acquired directly from Colonel Orr.
19 For more on this, see R.B. Byers, "Defence and Foreign Policy in the 1970s: The Trudeau Doctrine," *International Journal* 33 (1977-78).
20 "History of 423 Helicopter Squadron," http://www.shearwateraviationmuseum.ns.ca/.
21 This and the rest of the paragraph were retrieved from Glenn Cook, *Vignettes of a Canadian Naval Aviator, 1955-1983: A Memoir* (Ottawa: self-published, 2005), 266.
22 Peter Charlton, *Nobody Told Us It Couldn't Be Done: The Story of the VX10* (Ottawa: Peter Charlton, 1995), 155.
23 "History of 423 Helicopter Squadron," http://www.shearwateraviationmuseum.ns.ca/. The CO that year was Colonel (retired) Lee Myrhaugen.
24 Shadwick, "Replacing the *Sea King,*" 164-65.
25 Lynch, "Naval Shipborne Aircraft," 98; see also the hearings on the NSA/NSH project by the Standing Committee on National Defence and Veterans Affairs, 13 May 1993, Manson Papers, LAC, box 8, folder 6. As of March 2006, these papers were not yet available to the public, but special access was provided by General Manson.
26 Colonel (retired) Laurence McWha, e-mail to Aaron Plamondon, 5 March 2006.
27 NSA Project Management Office, "Definition Contract Proposals Evaluation Report," August 1987, included in a multi-volume package presented to Kim Campbell after she became Prime Minister; retrieved from the TASC consulting firm, Ottawa, briefing package 1, document 1, 1-2; see also Thomas Lynch, "New Shipborne Helicopter Program," *Wings* (1987-88): 98-101; and Thomas Lynch, "Canada's NSA Program: And Then There Was One," *Wings* (1988-89): 116-17.
28 See Chapter 2 of this volume on the first naval procurement SOR; see "Staff Requirements – RCN Helicopters," Appendix A to Minutes of the 492nd Meeting of the Naval Staff, 31 July 1951, DHH, 1000-100/3, 1-2.
29 Lynch, "New Shipborne Helicopter Program," 97.
30 NSA Project Management Office, "Definition Contract Proposals Evaluation Report," 1-1, 1-2.
31 Middlemiss, "Defence Procurement in Canada," 403-4.
32 James Fergusson, "In Search of a Strategy: The Evolution of Canadian Defence Industrial and Regional Benefits Policy," in Martin, *The Economics of Offsets,* 117-23.
33 Quoted in Ken Pole, "New Shipborne Aircraft – The Backbone of ASW," *Canadian Defence Quarterly,* Special Marketing Feature (December 1986): 6.
34 This and the next paragraph are based on the *White Paper on Defence, 1987,* 43, 89, 67.
35 Ibid., 78, 51, 75 respectively.
36 Douglas Bland, *The Administration of Defence Policy in Canada, 1947-85* (Kingston: R.P. Frye, 1987), 138.
37 NSA Project Management Office, "Definition Contract Proposals Evaluation Report," x, 1-2.

38 Westland Helicopters Ltd., "Aircraft Data Sheets, Rotor A-S," retrieved from Westland's official website, http://www.whl.co.uk/pdfs/history_rotor_a_s.pdf.
39 NSA Project Management Office, "Definition Contract Proposals Evaluation Report," x, 1-2.
40 See "EH-101 – More than Just an ASW Helicopter," *Defence* (1987), retrieved from "European Helicopter Industries," DHH, RA/CE.
41 NSA Project Management Office, "Aide Memoire on the New Shipborne Aircraft (NSA) Project," November 1991, TASC (consulting firm), briefing package 1, document 12, 12. The document was distributed to various ministries in November 1991 by the Ministry of Supply and Services to provide them with information on the status of the NSA project. See also Lynch, "New Shipborne Helicopter Program," 98-101; and Lynch, "Canada's NSA Program," 116-17.
42 James Fergusson, "In Search of a Strategy: The Evolution of Canadian Defence Industrial and Regional Benefits Policy," in Martin, *The Economics of Offsets,* 126.
43 Priestley, "Politics, Procurement Practices, and Procrastination," http://www.casr.ca/.
44 NSA Project Management Office, "Objectives of the NSA Project," TASC (consulting firm), briefing package 1, document 2, 29. For more on the research and development investments in Canadian industry, see Thomas Lynch, "Stuffing NSA: DND and Canadian Industry Gear Up to Provide Comprehensive Mission Suite," *Wings* (1987-88): 102-4.
45 NSA Project Management Office, "Aide Memoire on the New Shipborne Aircraft (NSA) Project," 12.
46 NSA Project Management Office, "Definition Contract Proposals Evaluation Report," 1-3; see also Robert Zincone, President, Sikorsky Aircraft Division, "Request for Proposal for New Shipborne Aircraft," letter to Gary Sanger, EHI Contract Manager, 4 February 1987, Manson Papers, LAC, box 8, folder 1.
47 Lynch, "Naval Shipborne Aircraft," 99-100.
48 This and the following paragraph were retrieved from NSA Project Management Office, "Definition Contract Proposals Evaluation Report," 1-3 to 1-5, 3-5, 3-16.
49 This and the following three lines were retrieved from A. Crosby, "Project Management in DND," *Canadian Defence Quarterly* 18, 6 Special no. 2 (1989): 59.
50 Michael Slack, "Canada's Defence Industrial Base: The Challenges," *Canadian Defence Quarterly* 18 (1989): 49.
51 This and the rest of the paragraph were retrieved from NSA Project Management Office, "Definition Contract Proposals Evaluation Report," v, vi, 2-5, 2-6, 2-11, 2-27, 3-2, 6-6, 8-1.
52 This and the next paragraph were retrieved from ibid., v, vi, 6-6, 6-18, 8-1.
53 NSA Project Management Office, "Aide Memoire on the New Shipborne Aircraft (NSA) Project," 12.
54 NSA Project Management Office, "NSA Alternative Aircraft Effectiveness," TASC, Campbell briefing package 1, document 2, 23, 30; see also Lynch, "New Shipborne Helicopter Program," 98-101; and Lynch, "Canada's NSA Program," 116-17.
55 Underwood, "The Eyes and Ears of NSA," 21.
56 Thomas Lynch, "Rescue at Sea: The Salvage of *Sea King* 12409," *Wings* (1988-89): 54.
57 NSA Project Management Office, "Aide Memoire on the New Shipborne Aircraft (NSA) Project," 13-14.
58 "*Sea Kings*' Days Should Be Cut Short," *Halifax Daily News,* 26 September 1989.
59 David Gibbings, Sea King*: 21 Years Service with the Royal Navy, 1969-1990* (Seaborough Hill, UK: Skylark Press, 1990), 15.
60 Hearings on the NSA/NSH by the Standing Committee on National Defence and Veterans Affairs, 13 May 1993, Manson Papers, LAC, box 8, folder 6.
61 "History of 423 Helicopter Squadron," http://www.shearwateraviationmuseum.ns.ca/.
62 Ibid.
63 Adcock, *H-3* Sea King *in Action,* 7.
64 Peter Charlton and Michael Whitby, *"Certified Serviceable"* Swordfish *to* Sea King*: The Technical Story of Canadian Naval Aviation by Those Who Made It So* (Gloucester, ON: CNATH, 1995), 423.
65 See the "History of 423 Helicopter Squadron," http://www.shearwateraviationmuseum.ns.ca/.

66 This and the following two sentences were retrieved from Charlton and Whitby, *"Certified Serviceable"* Swordfish *to* Sea King, 423, 426-27.
67 See Richard Gimblett, *Operation Friction: The Canadian Forces in the Persian Gulf, 1990-1991* (Toronto: Dundurn Press, 1997), 188-89, 70-72, 135, 142; see also "History of 423 Helicopter Squadron"; and Charlton and Whitby, *"Certified Serviceable"* Swordfish *to* Sea King, 424.
68 Maritime Command and Maritime Air Group Headquarters, "A New Shipborne Aircraft for Canada," 29 January 1993, DHH, 93/123, 5-6.
69 Colonel (retired) John Orr, e-mail to Aaron Plamondon, 12 December 2006.
70 NSA Project Management Office, "Aide Memoire on the New Shipborne Aircraft (NSA) Project," 13-14.
71 Ross Howard and Geoffrey York, "Helicopter Deal Up in the Air," *Globe and Mail,* 9 June 1992.
72 As explained by Commander Jake Freill at the Hearings on the NSA/NSH by the Standing Committee on National Defence and Veterans Affairs, 13 May 1993, Manson Papers, LAC, box 8, folder 6.
73 NSA Project Management Office, "Aide Memoire on the New Shipborne Aircraft (NSA) Project," 13-14.
74 Ibid.
75 See R.L. MacDonald, Director of Business for Boeing, "Request for Proposal," letter to Grant Johnson, Industrial Benefits Analysis Manager, 27 April 1990, Manson Papers, LAC, box 8, folder 1.
76 NSA Project Management Office, "NSA Alternative Aircraft Effectiveness," 23, 30; see also Lynch, "New Shipborne Helicopter Program," 98-101.
77 Hearings on the NSA/NSH by the Standing Committee on National Defence and Veterans Affairs, 13 May 1993, Manson Papers, LAC, box 8, folder 6.
78 The later model of the *EH-101,* the *Cormorant,* eventually fulfilled Canada's SAR role in 1998, five years after the NSH program was cancelled along with the NSA program.
79 See the statement made by R. Chisholm in British Columbia, *Official Report of the Debates in the Legislative Assembly,* 22 April 1993, Hansard 9, 5, 5365, http://www.leg.bc.ca/HANSARD/35th2nd/h0422pm.htm; see also "Another Reform Party Perspective on the Acquisition of a New Shipborne Aircraft and a New Search and Rescue Helicopter," Manson Papers, LAC, box 8, folder 7, 6.
80 NSA Project Management Office, "Aide Memoire on the New Shipborne Aircraft (NSA) Project," 17-18.

Chapter 6: The Vulnerability of the NSA

Epigraphs: *The Wednesday Report,* 24 February 1993, 1; Hearings on the NSA/NSH by the Standing Committee on National Defence and Veterans Affairs, 13 May 1993, Manson Papers, LAC, box 8, folder 6.

1 NSA Project Management Office, "Aide Memoire on the New Shipborne Aircraft (NSA) Project," November 1991, TASC, briefing package 1, document 12, 11.
2 See David Underwood, "The Eyes and Ears of NSA," *Canadian Aviation* (1987): 22.
3 Frank L. Boyd Jr., "The Politics of Canadian Defence Procurement: The New Fighter Aircraft Decision," in *Canada's Defence Industrial Base: The Political Economy of Preparedness and Procurement,* ed. David G. Haglund (Kingston: R.P. Frye, 1988), 142.
4 Dan Middlemiss, "Defence Procurement in Canada," in *Canada's International Security Policy,* ed. David B. Dewitt and David Leyton-Brown (Scarborough: Prentice Hall, 1995), 402.
5 Former ADM Mat Ray Sturgeon, interview with Aaron Plamondon, 20 February 2006.
6 Westland Helicopters Ltd., "Aircraft Data Sheets, Rotor A-S," retrieved from Westland's official website, http://www.whl.co.uk/pdfs/history_rotor_a_s.pdf.
7 Al Adcock, *H-3* Sea King *in Action* (Corrollton, TX: Signal Publications, 1995), 49.
8 Department of National Defence, "NSA/NSH Industrial and Regional Benefits," March 1993, TASC, briefing package 1, document 6, 2. See also Department of National Defence, "IRB Offer – Regional Benefits," TASC, briefing package 1, document 2, 31.
9 "Quebec to Receive Lion's Share of $4.4-Billion Helicopter Deal," *Globe and Mail,* 25 July 1993.

10 *Canada Defence Policy 1992,* 13. Retrieved from the DND webpage at http://www.forces.gc.ca/admpol/newsite/downloads/CanadaDefPolE_all.pdf.
11 Danford Middlemiss, "Canadian Defence Funding: Heading towards Crisis?" *Canadian Defence Quarterly* 21 (1991): 13.
12 Robert Bothwell and J.L. Granatstein, *Our Century: The Canadian Journey in the Twentieth Century* (Toronto: McArthur, 2000), 226.
13 Peter C. Newman, *The Secret Mulroney Tapes: Unguarded Confessions of a Prime Minister* (Toronto: Random House Canada, 2005), 13.
14 "Quebec to Receive Lion's Share of $4.4-Billion Helicopter Deal," *Globe and Mail,* 25 July 1993.
15 J.W. MacAleese, "The CH-146 *Griffon:* Underrated and Over Criticized," paper written at the Canadian Forces College, 2001, http://wps.cfc.forces.gc.ca/.
16 See Chapter 1 in this volume.
17 The section on the *CF-18* maintenance contract was retrieved from Middlemiss, "Defence Procurement in Canada," 402-3; for more on the selection process of the *CF-18,* see Paul Manson, "The CF-18 *Hornet*: Canada's New Fighter Aircraft," *Canadian Defence Quarterly* 10 (1980).
18 Michel Rossignol, "Notes on the Purchase of the EH-101 Helicopters," 7 June 1993, Library of Parliament, Research Branch, Political and Social Affairs Division, TASC, briefing package 2, document 1, 7.
19 David Pugliese, "400 Firms Will Take Part in Helicopter Project," Montreal *Gazette,* 25 July 1992.
20 Rossignol, "Notes on the Purchase of the EH-101 Helicopters," 4.
21 "Another Reform Party Perspective on the Acquisition of a New Shipborne Aircraft and a New Search and Rescue Helicopter," Manson Papers, LAC, box 8, folder 7, 8.
22 David Pugliese and M. Kennedy, "The Hunt for a Helicopter," *Ottawa Citizen,* 4 July 1992, B2.
23 David Godfrey, "Procuring Canada's New Helicopters: Still Firmly on the Rails, the Canadian Navy's New Shipborne Aircraft Program Has Survived Severe Cutbacks in Defence Spending," *Wings* (1991-92): 38.
24 "Maritime Operations and the Shipborne Helicopter," [1992], TASC, briefing package 1 of 2, document 2, 3-5.
25 Department of National Defence, "DND News Release – Communiqué," 24 July 1992, retrieved from "European Helicopter Industries," DHH, RA/CE.
26 Maritime Command and Maritime Air Group Headquarters, "A New Shipborne Aircraft for Canada," 29 January 1993, DHH, 93/123, 2-3.
27 "Maritime Operations and the Shipborne Helicopter," 3-5.
28 This and the rest of the paragraph were retrieved from ibid., 6.
29 This and the rest of the paragraph were retrieved from ibid., 7.
30 Ibid., 8.
31 See the "History of 423 Helicopter Squadron," http://www.shearwateraviationmuseum.ns.ca/.
32 Maritime Command and Maritime Air Group Headquarters, "A New Shipborne Aircraft for Canada," 5-6.
33 "Maritime Operations and the Shipborne Helicopter," 8.
34 Department of National Defence, "NSA/NSH Defence Policy Basis," July 1992, retrieved from TASC, briefing package 1, document 6, 2.
35 "Maritime Operations and the Shipborne Helicopter," 17.
36 This and the rest of the paragraph were retrieved from "NSA – Present Fleet's Minimum Option," TASC (consulting firm), briefing package 1, document 2, 28.
37 "NSA Alternative Aircraft Effectiveness," [1992], TASC, briefing package 1, document 2, 23, 43.
38 "EH-101 Remains the Best Choice for DND," TASC, briefing package 1, document 2, 30.
39 Jean Chrétien to Paul Manson, 2 July 1992, Manson Papers, LAC, box 8, folder 3, 2.
40 Paul Manson to Jean Chrétien, 13 July 1992, Manson Papers, LAC, box 8, folder 3, 1-2.

41 Press release by the Office of the Opposition, House of Commons, 24 July 1992, Manson Papers, LAC, box 8, folder 3, 30.
42 This and the following quotation were retrieved from letter to Peace Alliance Supporters, 2 September 1992, Manson Papers, LAC, box 8, folder 3.
43 "Notes for a Speech by the Honourable Jean Chrétien to the University of Ottawa Law Faculty," 27 January 1993, Manson Papers, LAC, box 8, folder 5, 8.
44 "Briefing Note on Remarks by Honourable Jean Chrétien, Leader of the Liberal Party, NSA/NSH Program in Canada," Manson Papers, LAC, box 8, folder 5.
45 Canada, House of Commons, *Debates,* 25 February 1993, retrieved from Manson Papers, LAC, box 8, folder 5.
46 Hearings on the NSA/NSH by the Standing Committee on National Defence and Veterans Affairs, 13 May 1993, Manson Papers, LAC, box 8, folder 6.
47 This and the rest of the paragraph were retrieved from "High Tech Helicopters Called Expensive Necessity," n.d., retrieved from Manson Papers, LAC, box 8, folder 5.
48 Robert Zincone, President, Sikorsky Aircraft Division, "Request for Proposal for New Shipborne Aircraft," letter to Gary Sanger, EHI Contract Manager, 4 February 1987, Manson Papers, LAC, box 8, folder 1.
49 Rossignol, "Notes on the Purchase of the EH-101 Helicopters," 6.
50 "Director General Public Affairs – Questions and Answers – NSA/NSH – General," 18 February 1993, TASC, briefing package 1, document 5, 4, 8; see also Maritime Command and Maritime Air Group Headquarters, "A New Shipborne Aircraft for Canada," 4; for more on the continued submarine threat in the post-Cold War era, see Peter Haydon and Don Macnamara, "Of Helicopters and Defence Policy," Canadian Institute of Strategic Studies Datalink 40, March 1993.
51 "Director General Public Affairs – Questions and Answers – NSA/NSH – General," 2, 9.
52 Ibid., 4, 8.
53 This and the rest of the paragraph were retrieved from ibid., 5-6.

Chapter 7: The 1993 NSA Cancellation

Epigraphs: Maritime Command and Maritime Air Group Headquarters, "A New Shipborne Aircraft for Canada," 29 January 1993, DHH, 93/123, 3-4; Jim Brown, "Full Circle: Electoral Promise to Axe Choppers Returns to Haunt Chrétien," Canadian Press Newswire, 5 January 1998; Paul Manson, "Procurement Cycle Growth: The Race between Obsolescence and Acquisition of Military Equipment in Canada, 1960 to the Present," paper presented at the Canadian Institute of Strategic Studies Seminar, Toronto, 22 July 2005.

1 This and the rest of the paragraph were retrieved from Michel Rossignol, "Costs of Cancelling EH-101 Contracts," 15 February 1993, Political and Social Affairs Division, Research Branch, Library of Parliament, TASC, briefing package 2, document 3.
2 Canada, House of Commons, *Minutes of Proceedings and Evidence of the Standing Committee on National Defence and Veterans Affairs,* 13 May 1993, 48, 71.
3 Michel Rossignol, "Notes on the Purchase of the EH-101 Helicopters," 7 June 1993, Library of Parliament, Research Branch, Political and Social Affairs Division, TASC (consulting firm), briefing package 2, document 1, 7.
4 John Brewin, "Defence Is Splurging on Dividends of Peace," *Ottawa Citizen,* 27 July 1992.
5 Paul Manson to Minister, 13 July 1992, Manson Papers, LAC, box 8, folder 3, 1.
6 David Pugliese, "A Whole New Whirled of Trouble," *Ottawa Citizen,* n.d., retrieved from "European Helicopter Industries," DHH, RA/CE.
7 Colonel (retired) Laurence McWha, interview with Aaron Plamondon, 10 February 2006.
8 Norman Hillmer and J.L. Granatstein, *Empire to Umpire: Canada and the World to the 1990s* (Toronto: Copp Clark Longman, 1994), 344-45.
9 John Ward, "Critics of New Helicopters Like Them, but Price Tag Gives Them Ammunition," Montreal *Gazette,* 11 March 1993.
10 Peter Charlton and Michael Whitby, *"Certified Serviceable"* Swordfish *to* Sea King*: The Technical Story of Canadian Naval Aviation by Those Who Made It So* (Gloucester, ON: CNATH, 1995), 389.

11 Hearings of the NSA/NSH by the Standing Committee on National Defence and Veterans Affairs, 29 April 1993, Manson Papers, LAC, box 6, folder 8.
12 This and the following quotation are from Dale Grant, ibid., 13 May 1993.
13 This and the following paragraph were retrieved from ibid., 29 April 1993, box 5, folder 8.
14 This and the following paragraph were retrieved from "Speaking Notes for the Honourable Kim Campbell, Minister of National Defence and Minister of Veterans Affairs to the Standing Committee on National Defence and Veterans Affairs," 5 May 1993, Manson Papers, LAC, box 8, folder 6, 1-3; see also ibid., 5 and 13 May 1993, for the actual meeting notes of the committee.
15 For this quotation and the rest of the paragraph, see ibid.
16 *Creating Opportunity: The Liberal Plan for Canada,* retrieved from TASC bundle 32.
17 Hearings on the NSA/NSH by the Standing Committee on National Defence and Veterans Affairs, 13 May 1993, Manson Papers, LAC, box 8, folder 6.
18 As discussed at ibid., 29 April 1993, box 5, folder 8.
19 Colonel (retired) Laurence McWha, e-mail to Aaron Plamondon, December 2004.
20 "Briefing Note for Minister of National Defence on Cancellation of the NSA/NSH Project," 27 October 1993, TASC, bundle 32, 2.
21 Jeff Sallot and Ross Howard, "PM Chops Seven Helicopters," *Globe and Mail,* 3 September 1993.
22 Meeting of the Standing Committee on National Defence and Veterans Affairs, 5 May 1993, Manson Papers, LAC, box 8, folder 6.
23 Terrence Willis, "Tories Cut Helicopter Purchase to 43," Montreal *Gazette,* 3 September 1993.
24 Michel Gratton, "Choppy Going," Ottawa *Sunday Sun,* 5 September 1993.
25 Department of National Defence, "EH-101 Helicopter: Budget Reduction,"September 1993, retrieved from "European Helicopter Industries," DHH, RA/CE.
26 "Another Campbell Cop Out," *Ottawa Citizen,* 4 September 1993.
27 "Campbell Takes Flak for Cutting Helicopters," *Ottawa Citizen,* 3 September 1993.
28 Review of NSA/NSH by the Standing Committee on National Defence and Veterans Affairs, 29 April 1993, Manson Papers, box 5, folder 8.
29 Hugh Winsor, "PM's Tour Bus Springs a Leak," [September 1993], Parliamentary Bureau, Ottawa, retrieved from "European Helicopter Industries," DHH, RA/CE; the French press was equally negative concerning the political move of reducing the NSA/NSH program. See Chantal Hébert, "Une gigantesque erreur de calcul," *Le Devoir,* 5 September 1993.
30 Sharon Hobson, "Shipbuilding and Defence," *Canadian Sailings,* 27 September 1993, retrieved from "European Helicopter Industries," DHH, RA/CE.
31 Dale Grant, "Reducing Numbers Won't Make EH-101 Fly," *Toronto Star,* 6 September 1993.
32 "Kim Campbell and Her Helicopters," *Globe and Mail,* 11 March 1993.
33 This and the following quotation were retrieved from Geoffrey York, "Tory Prospects Concerned EH-101 Makers," *Globe and Mail,* 22 October 1993.
34 H.P. Neilsen, "Transition Package in the Event of a Liberal Majority," 18 October 1993, TASC (consulting firm), bundle 32, 4-8.
35 H.P. Nielsen, "Briefing Note for Minister of National Defence on Cancellation of the NSA/NSH Project," 19 October 1993, TASC, bundle 32, 1-3.
36 "Briefing Note for Minister of National Defence on Cancellation of the NSA/NSH Project," 27 October 1993, TASC, bundle 32, 2.
37 This and the following quotation were retrieved from John Cody, "Colonel Cody Replies," *Warrior* 22, 9 (4 May 1995).
38 The information on operations in the former Yugoslavia and Haiti was retrieved from "History of 423 Helicopter Squadron," http://www.shearwateraviationmuseum.ns.ca/.
39 Harvey Nielsen, "Briefing Note to the Minister on the EH-101 Helicopter Project," 27 October 1993, TASC, bundle 32, 3.
40 This and the rest of the paragraph were retrieved from e-mail correspondence with Colonel (retired) Laurence McWha, 12 March 2006.
41 Ian Austen, "No Compensation on Copters, Quebec Told," Montreal *Gazette,* 5 November 1993.

42 This and the following quotation were retrieved from Geoffrey York, "Helicopter Deal Cancelled without Regret," *Globe and Mail,* 5 November 1993; see also Brian Underhill, "Chrétien Downs Helicopters," Halifax *Chronicle-Herald,* 5 November 1993.
43 Article F4 of the NSA/NSH contract and the following quotation were received from former ADM Mat, NDHQ, Ray Sturgeon during an interview on 20 February 2006.
44 Ibid.
45 General (retired) Paul Manson, interview with Aaron Plamondon, 24 September 2005.
46 General (retired) George Macdonald, former Director General Aerospace Developmnent, NDHQ, interview with Aaron Plamondon, 25 September 2005.
47 Sharon Hobson, "Will Liberals Attack Defence Policy?" *The Financial Post,* 4 November 1993.
48 Retrieved from the official website of the Canadian Department of National Defence, http://www.dnd.ca/somalia/vol1/v1c6e.htm.
49 Douglas Bland, "A Unified Theory of Civil Military Relations," *Armed Forces and Society* 26, 1 (1999): 9.
50 Paul Manson, "Opening Remarks by P.D. Manson to the Standing Committee on National Defence and Veterans Affairs," 13 May 1993, Manson Papers, LAC, box 8, folder 6.

Chapter 8: The 1994 White Paper and the New Statement of Requirement

Epigraph: Based on "Statement of Capability Deficiency," Maritime Helicopter Project, June 1994, an updated version of the original statement,which predated the November 1993 contract cancellation. See Department of National Defence, "Statement of Operational Requirement – Final," 14 July 1999, retrieved from http://www.forces.gc.ca/.

1 This and the rest of the paragraph were retrieved from John Yorston, "It's Too Late to Protest Chopper Cuts, Writers Say," Montreal *Gazette,* 5 November 1993.
2 Don Macpherson, "Waking Up: Quebecers Finally Quit Yawning over Killing of Helicopters," Montreal *Gazette,* 4 November 1993.
3 Ian Austen, "No Compensation on Copters, Quebec Told," Montreal *Gazette,* 5 November 1993.
4 "Putting the Money to Better Use," Montreal *Gazette,* 4 November 1993.
5 Austen, "No Compensation on Copters."
6 General (retired) Paul Manson, "Procurement Cycle Growth: The Race between Obsolescence and Acquisition of Military Equipment in Canada, 1960 to the Present," paper presented to the Canadian Institute of Strategic Studies Seminar, Toronto, 22 July 2005.
7 Austen, "No Compensation on Copters."
8 This and the rest of the paragraph were retrieved from David Collenette, "Budget Impact: National Defence," February 1994, acquired from former ADM Mat Ray Sturgeon, 1-3, 5, 12.
9 David Pugliese, "What's Riding on the Chopper?," *Ottawa Citizen,* 20 April 2002.
10 "For the Special Joint Committee on the Review of Canada's Defence Policy," 13 July 1994, TASC, briefing package 2, document 2, 1/6. Annex D includes the category of "Accidents and Forced Landings."
11 Colonel (retired) John Cody, e-mail to Aaron Plamondon, 12 January 2006.
12 Department of National Defence, *White Paper on Defence, 1994* (Ottawa: Canada Communication Group, 1994), 46.
13 Department of National Defence, "Statement of Requirement, Maritime Helicopter," 25 October 1994, TASC (consulting firm), bundle 32, file 32680-304.
14 This and the rest of the paragraph were retrieved from Rear Admiral G.L. Garnett, "Maritime Helicopter SOR," TASC, bundle 32, 1.
15 "Maritime Operations and the Shipborne Helicopter," [1992], TASC, briefing package 1, document 2, 3-5.
16 Garnett, "Maritime Helicopter SOR," 1.
17 Colonel (retired) John Orr, "Forty Years On: The *Sea King* Saga," *Air Force* (2003): 7; electronic copy given to Aaron Plamondon.
18 This and the rest of the paragraph were retrieved from the memo by Lieutenant Colonel J.F. Cottingham, 7 February 1995, TASC, bundle 32.

19 This and the following quotation were retrieved from a letter from Major A.D. Blair to unknown officer, 9 February 1995, TASC, bundle 32.
20 Memorandum on SOR to MARCOM, February 1995, obtained under Access to Information Act (AIA) (TASC file AGU016453).
21 Comments on SOR, February 1995, AIA (AGU021297).
22 Stuart E. Soward, *Hands to Flying Stations: A Recollective History of Canadian Naval Aviation*, vol. 2 (Victoria: Neptune Developments, 1995), 462.
23 Ibid., 461.
24 This and the rest of the paragraph were retrieved from "Statement of Operational Requirement (SOR) – NSA/MHP," December 1995, TASC, bundle 8, A-1.
25 This and the rest of the paragraph were retrieved from "Details of the Operational Requirement," TASC, bundle 8.
26 This and the rest of the paragraph were retrieved from "Statement of Operational Requirement," D-1, D-2.
27 PWGSC, "Settlement Reached with EH Industries for the EH-101 Helicopter Program," news release, 23 January 1996.
28 Department of National Defence, *1994-95 Main Estimates, DND Part III Expenditure Plan.*
29 John Geddes, "Helicopter Termination Cost Less than Ottawa Expected," *Financial Post*, 24 January 1996.
30 Former ADM Mat Ray Sturgeon, e-mail to Aaron Plamondon, 24 July 2006.
31 *The Wednesday Report*, 12 May 1993, 5; 20 October 1993, 4.
32 *The Wednesday Report*, 21 April 1993, 5. The initial order, valued at $8.9 million, called for 100 of the composite landing gear covers with the first 44 going to the United Kingdom.
33 *The Wednesday Report*, 20 October 1993, 6.
34 Michel Rossignol, "*Sea King* and *Labrador* Helicopters in the Canadian Forces: Current Status and Possible Options," Research Branch, Library of Parliament, 11 July 1994, retrieved from TASC (consulting firm), briefing package 1, document 11, 21.
35 David Miller, "Extending ASW: Helicopters for Destroyers and Frigates," *International Defense Review* (1993): 894.
36 Geddes, "Helicopter Termination Cost Less than Ottawa Expected."
37 "Book Closes on EH101 Deal, Coffers Open," *Globe and Mail*, 10 November 1995.
38 David Collenette to Malcolm Rifkind, 4 May 1995, and to Domenico Corcione, 13 September 1995; letters received from AWIL directly.
39 This and the rest of the paragraph were retrieved from "Briefing Note: Maritime Helicopter Project"; acquired from Colonel (retired) Laurence McWha.
40 PWGSC, *News Release*, 8 November 1995.
41 CSH Statement of Interest, 29 January 1996.
42 See the KPMG proposal, 17 July 1997, AIA (AGU066618), and the KPMG contract (W8475-HF12/001/ZG), 25 July 1997, AIA (AGU071380).
43 KPMG, *Independent Validation and Verification of Bid Evaluations for the CSH Project*, August 1997, AIA (AGU016782).
44 Ellen Stensholt to Charles Dubin, February 1998, AIA (AGU023266); Charles Dubin to Ellen Stensholt, 27 February 1998, AIA (AGU023258).
45 Pierre Lagueux, e-mail to Vice Admiral G.L. Garnett, 20 January 1998, AIA (AGU021936).
46 This and the rest of the paragraph were retrieved from "Maritime Helicopter Project, SOR Review," briefing to VCDS, ADM Mat, Chief of Air Staff (CAS), and Chief of Maritime Staff (CMS), 9 February 1998, AIA (AGU003106).
47 E.J. Lerhe, Canadian Forces Maritime Warfare Centre, "MH SOR Review – Endurance Requirement in the Littoral Environment," 23 April 1998, AIA (AGU036736).
48 This and the rest of the paragraph were retrieved from "MHP SOR Briefing to CAS and CMS," 22 June 1998, AIA (AU002970).
49 Pierre Lagueux, e-mail to Vice Admiral G.L. Garnett, 7 December 1998, AIA (AGU016448).
50 "Maritime Helicopter Project – Potential Contenders," 15 February 1999, AIA (AGU029445).
51 See "MHP SOR Briefing to Minister of National Defence, 12 July 1999," AIA (AGU073259).
52 Lieutenant General David Kinsman, e-mail to Colonel G. Sharpe, 16 May 1999, AIA (AGU071600).

53 See Colonel Brian Akitt, "The *Sea King* Replacement Project: A Lesson in Failed Civil-Military Relations," Canadian Forces College, 2003, 23; paper obtained from Colonel Akitt.
54 "MHP SOR Briefing to Minister of National Defence, 12 July 1999," AIA (AGU073259).
55 PWGSC, "Maritime Helicopter Project: Procurement Strategies," 16 November 1999, AIA (AGU050696).
56 EHI to Alan Williams, 7 December 1999, AIA (AGU002030).
57 "*Sea King* Successfully Recovered after Landing on Water," 2 December 1999, http://www.dnd.ca/.
58 "Lack of Money: The Root of All Evil?" Military Affairs and Defence Committee (Royal Canadian Military Institute, 2000), 3.

Chapter 9: The Maritime Helicopter Project

Epigraph: SCONDVA, "Procurement Study," 14 June 2000.

1 Mike Blanchfield, "Government Not Hiding Helicopter Cost," *National Post,* 25 August 2001.
2 "Briefing Note – the Maritime Helicopter Project," 3 May 2001,obtained by Colonel (retired) Laurence McWha, 12; see also David Rattray, Assistant Auditor General, to Jane Billings, Deputy Minister of PWGSC, 6 April 2001, AIA (AGU025239).
3 Briefing on the MHP by Deputy Minister of PWGSC to Minister of PWGSC, 9 March 2001, AIA (AGU045110).
4 G. Bologna to Art Eggleton regarding the maritime helicopter procurement strategy, 2 October 2000, AIA (AGU002003).
5 Colonel Lee Myrhaugen, "Maritime Helicopter Procurement Process," http://www.naval.ca/.
6 Federal Court of Appeal determination, 7 March 2001, retrieved from Paul M. Lalonde, "Hurray, We Lose," *Summit* 4, 2 (June 2001).
7 "Helicopter Lobbyist Joins PM's Office," *Ottawa Citizen,* 30 March 2001.
8 Sheila Fraser, Auditor General, to Senator Michael Forestall, 6 April 2001, AIA (AGU001990).
9 David Rattray, Assistant Auditor General, to Jane Billings, Deputy Minister of PWGSC, 6 April 2001, AIA (AGU025239).
10 Raymond Chrétien, "Renewed French Misgivings," e-mail, 3 April 2001, AIA (AGU034096).
11 This and the rest of the paragraph were retrieved from an e-mail from Major N.B. Barrett to N. Crawley and Colonel Drummond at the MHP PMO, 11 April 2001, AIA (AGU039876).
12 Colonel Henneberry, memorandum, 4 May 2001, AIA (AGU024505).
13 SCONDVA transcript, 5 June 2001.
14 Canada, House of Commons, Question Period transcript, 6, 7, 12 June 2001.
15 Colonel Istchenko to Colonel Hinke at 12 Wing, Shearwater, 27 July 2001, AIA (AGU027463); for a list of the proposed reductions, see "MHP Proposed Changes," AIA (AGU027464).
16 Stephanie Rubec, "Bidder Gives Up on Chopper Contract," http://forums.army.ca/.
17 "U.S. Giant Aims to Win Copter Deal," *Ottawa Citizen,* 11 June 2003.
18 Hugh Winsor, "Common Sense Triumphs in Helicopter Saga," *Globe and Mail,* 6 December 2002.
19 John Manley, interview with Don Newman, *Politics,* CBC Newsworld, 1 May 2003.
20 Paul Martin, comments at a Town Hall Meeting organized by the Centre for Foreign Policy Studies, Halifax, 15 May 2003.
21 David J. Bercuson and Barry Cooper, "Helicopter Replacement Fiasco," *Fraser Forum* (June 2003), http://www.fraserinstitute.org/Commerce.Web/product_files/Helicopter%20Replacement%20Fiasco-cooper0603.pdf.
22 Peter O'Neil and Mike Blanchfield, "Chopper Bidder's Tactics Anger Rivals," *Ottawa Citizen,* 5 June 2003.
23 Colonel Brian Akitt, "The *Sea King* Replacement Project: A Lesson in Failed Civil-Military Relations," Canadian Forces College, 2003, 9; paper obtained from Colonel Akitt.
24 Ibid., 9; Samuel Huntington, *The Soldier and the State: The Theory and Politics of Civil-Military Relations* (Cambridge, MA: Belknap Press, 1964), 95.
25 See also Douglas Bland, "A Unified Theory of Civil Military Relations," *Armed Forces and Society* 26, 1 (1999).
26 Akitt, "The *Sea King* Replacement Project," 20-22.

27 Advisory Committee on Administrative Efficiency, "Achieving Administrative Efficiency," report to the Minister of National Defence, vii, http://www.dnd.ca/site/Focus/AE/AEReportFull_e.pdf.
28 John McCallum, speech to the Canadian Defence Industries Association, Ottawa, 22 October 2003.
29 "Ottawa in No Hurry for New Choppers," Halifax *Chronicle-Herald,* 26 May 2003.
30 Giovanni de Briganti, "Canada's Continuing Heli-Saga," *Defence Daily Network,* 1 July 1999.
31 Auditor General of Canada, "National Defence: In Service Equipment," report to the House of Commons, December 2001, http://www.oag-bvg.gc.ca/internet/English/osh_20020221_e_23337.html.
32 "Aircraft Occurence Summary," for *Sea King* 124A422, http://www.airforce.forces.gc.ca/dfs/reports-rapports/I/pdf/fti/ch12422.pdf.
33 David Pugliese, "What's Riding on the Chopper?" *Ottawa Citizen,* 20 April 2002.
34 Mike Blanchfield, "*Sea King* Crew Used Cellphone to Get Help," *Ottawa Citizen,* 29 April 2001.
35 "*Sea King* Helicopter Makes Emergency Landing," *Globe and Mail,* 4 September 2002.
36 "Sea King Crash Will Delay Gulf Mission," CBC News Online, http://www.cbc.ca/news/.
37 Canadian Press, "Canadian Military Short of *Sea King* Mechanics," http://www.ctv.ca.
38 Sharon Hobson, "Plain Talk: The Process of (Not) Acquiring Maritime Helicopters," *Canadian Naval Review* 4, 4 (2009): 40.
39 "Maritime Helicopter Project – Prequalification Process," http://www.dnd.ca/.
40 Federal Court of Canada, "Federal Court between AgustaWestland International Limited and Minister of Public Works and Government Services and Sikorsky International Operations, Inc.," 1 September 2004, court file T-1605-04, 9.
41 Hobson, "The Process of (Not) Acquiring Maritime Helicopters," 40.
42 These details were recounted by Colonel (retired) Laurence McWha, part of the *Cormorant* team during the competition, e-mail to Aaron Plamondon, 15 February 2008.
43 Pugliese, "What's Riding on the Chopper?"
44 "Will the *Sea Kings* Still Be in Service in 2011 and Beyond?," *Hill Times,* 19 July 2004.
45 See the *Hill Times,* 20 October and 10 November 2003.
46 Member of Team *Cormorant* speaking on the basis of anonymity.
47 David J. Bercuson, Aaron P. Plamondon, and Ray Szeto, *An Opaque Window: An Overview of Some Commitments Made by the Government of Canada Regarding the Department of National Defence and the Canadian Forces* (Calgary: Canadian Defence and Foreign Affairs Institute, 2006), 28.
48 "On Your Wing: MHP Taking Off!" retrieved from L-3's official website, http://www.mas.l-3com.com/doc/OYW/OnYourWing0003_eng.pdf.
49 Department of National Defence, "Backgrounder," 23 November 2004, http://www.forces.gc.ca/.
50 Retrieved from GDC's official website at http://www.gdcanada.com/documents/2007_AN_brochure1.pdf.
51 This was retrieved from the Department of National Defence, "Briefing Re: Maritime Helicopter Project at NDHQ," 23 July 2004, transcript prepared by Media Q exclusively for the Department of National Defence, AIA (AGU007940), 3-10, 15.
52 Department of National Defence, "Backgrounder," 23 November 2004, http://www.forces.gc.ca/.
53 This and the following paragraph are based on the Department of National Defence, "Briefing Re: Maritime Helicopter Project at NDHQ," 3-10, 15.
54 Colonel (retired) Laurence McWha, e-mail to Aaron Plamondon, 15 February 2008.
55 Gloria Galloway, "Winning Helicopters Assailed as Inferior," *Globe and Mail,* 24 July 2004.
56 PWGSC, "Solicitation Amendment of Liquidated Damages," 18 December 2003, Solicitation W847001MP01/C, paragraph 6.4.1.
57 Department of National Defence, "Briefing Re: Maritime Helicopter Project at NDHQ," 4; see also "Date Given for Delivery Not Feasible, Critics Say," *Globe and Mail,* 24 July 2004.

58 David Pugliese, "Defence Report Gives Losing Chopper Firm Fuel for Lawsuit: 1997 Review Favoured Same Craft for Search, Naval Missions," *Ottawa Citizen,* 24 July 2004.
59 This and the following quotation were retrieved from Aaron Plamondon, "Heli-Hijinks," *Calgary Herald,* 29 July 2004.
60 "Date Given for Delivery Not Feasible, Critics Say," *Globe and Mail,* 24 July 2004.
61 This and the following paragraph were retrieved from Federal Court of Canada, "Federal Court between AgustaWestland International Limited and Minister of Public Works and Government Services and Sikorsky International Operations, Inc.," 6, 3, 10-14.
62 Michael Hutton, Canadian Press, "Court Case Going Ahead over Ottawa's Choice for $5-Billion Helicopter Deal," 6 April 2005.

Chapter 10: The *Cyclone* Decision

Epigraph: Interview with Dan Ross, *Canadian Defence Review* 13, 2 (2007): 47.

1 Department of National Defence, obtained under Access to Information Act (AIA) Request A-2005-01101.
2 Sharon Hobson, "Winds of Change: *Cyclone* Brings Canada Up to Date," *Jane's Navy International* (2005): 28.
3 AWIL, Amended Statement of Claim, April 2006, received from Will Macdonald of AWIL.
4 *Knight-Ridder Tribune,* 22 February 2006.
5 Halifax *Chronicle-Herald,* 24 January 2007.
6 This and the following sentence were retrieved from David Pugliese, "The Most Overlooked Defence Story of 2007? (Or Why a $1 Billion Lawsuit over the Maritime Helicopter Program Went Away So Quietly)," *Ottawa Citizen,* 2 January 2008.
7 PWGSC, "The Government of Canada Announces Settlement with AgustaWestland International Limited," press release, 26 November 2007, http://www.tpsgc-pwgsc.gc.ca/.
8 Joe Schneider, "Finmeccanica Drops Helicopter Suit against Canada," http://www.bloomberg.com/.
9 PWGSC, "The Government of Canada Announces Settlement."
10 See Chapter 8 of this volume.
11 Paul M. Lalonde, "Edmonton Case Splits Supreme Court," *Summit* 10, 4, (2007).
12 Colonel (retired) Laurence McWha, e-mail to Aaron Plamondon, 11 February 2008.
13 This and the following quotation were retrieved from Richard Foot, "No *Sea King* Successor until 2010," *Ottawa Citizen,* 10 January 2008.
14 This and the following quotation were retrieved from Steve Rennie, "New Copters Delayed," Halifax *Chronicle-Herald,* 10 January 2008.
15 Mike Blanchfield, "Callous Liberals to Blame in Chopper Delays," *Ottawa Citizen,* 12 January 2008.
16 Rennie, "New Copters Delayed."
17 Dan Ross, "How to Buy a Tank," *Calgary Herald,* 7 February 2008, in direct response to Aaron Plamondon, "Buying Military Equipment Still a Tricky Venture," *Calgary Herald,* 17 January 2008.
18 Sharon Hobson, "Plain Talk: The Process of (Not) Acquiring Maritime Helicopters," *Canadian Naval Review* 4, 4 (2009): 40.
19 Directorate of Flight Safety, "Crash of *Sea King* Helicopter CH124-38 on February 2, 2006," at http://www.airforce.forces.gc.ca/dfs/home-accueil-eng.asp.
20 Canadian Press, "Military Studying Extending Lifespan of *Sea Kings,*"14 May 2008, http://montreal.ctv.ca/servlet/an/plocal/CTVNews/20080514/seakings_extension_080514/20080514/?hub=OttawaHome.
21 Daniel Leblanc and Steven Chase, "$117-Million Later, Ottawa's Troubles with Sikorsky Aren't Over," *Globe and Mail,* 7 January 2009.
22 Daniel Leblanc, "Ottawa Withholding Sikorsky Payments," *Globe and Mail,* 14 May 2008.
23 David Pugliese, "No Delay at All on *Cyclone* Helicopters," *Ottawa Citizen,* 25 June 2008, http://communities.canada.com/ottawacitizen/.
24 This and the rest of the paragraph were retrieved from Michael Tutton, "Cost of Choppers Rising Defence Minister Says," *Toronto Star,* 18 September 2008.

25 Mike Blanchfield, "MacKay Blasts Chrétien over Military Helicopter Delay," Regina *Leader-Post,* 12 January 2008.
26 The announcement came in Gatineau, Quebec, and was reproduced, along with the DND backgrounder, on the DND web page. This and the rest of the paragraph were retrieved from http://news.gc.ca/.
27 Hobson, "Plain Talk: The Process of (Not) Acquiring Maritime Helicopters," 40.
28 This and the following two sentences were retrieved from David Pugliese, "New Engines for the Troubled *Cyclone* Helicopter?" *Ottawa Citizen,* 20 February 2009, http://communities.canada.com/ottawacitizen/.
29 Sharon Hobson, "Canada Amends Its Sikorsky *Cyclone* Contract," *Jane's Defence Weekly,* 30 December 2008.
30 This and the following quotation were retrieved from David Pugliese, "Why the Secrecy of the *Cyclone* Helicopter Changes?" *Ottawa Citizen,* 12 December 2008, http://communities.canada.com/ottawacitizen/. For how this had generally been the case at the DND since 2006, see David J. Bercuson, Aaron P. Plamondon, and Ray Szeto, *An Opaque Window: An Overview of Some Commitments Made by the Government of Canada Regarding the Department of National Defence and the Canadian Forces* (Calgary: Canadian Defence and Foreign Affairs Institute, 2006).
31 Department of National Defence, "Briefing Re: Maritime Helicopter Project at NDHQ," 23 July 2004, transcript prepared by Media Q exclusively for the DND, AIA (AGU007940), 3-10, 15.
32 Blanchfield, "MacKay Blasts Chrétien."
33 This and the rest of the paragraph were retrieved from Sharon Hobson, "S-92 Crash Raises Questions about Canada's H-92 Purchase," *Jane's Defence Weekly,* 9 April 2009.
34 Peter Cheney, "S-92 CONTRACT Sikorsky Objects to Claims Helicopter Failed to Achieve Certification," *Globe and Mail,* 8 April 2009.
35 Peter Cheney, "Lawsuit Claims Sikorsky Knew S-92 Had Fatal Flaw, but Failed to Inform Pilots," *Globe and Mail,* 20 June 2009.
36 This paragraph and the following three paragraphs were retrieved from "Janet Breen, on her own behalf and on behalf of the estate of Peter Breen, et al. vs. Keystone Helicopter Corporation, et al.," Court of Common Pleas, Philadelphia County, Case ID: 090601841. Retrieved from http://beta.images.theglobeandmail.com/archive/00079/Sikorsky_suit_79387a.pdf.
37 The FAR 29 defines remote as one flight hour in ten million. The lawsuit claims that Sikorsky was experiencing complete oil loss approximately every 50,000 hours.
38 This quote and the rest of the paragraph were retrieved from Sharon Hobson, "S-92 Crash Raises Questions about Canada's H-92 Purchase," *Jane's Defence Weekly,* 9 April 2009.
39 Allison Auld, "Choppers Must Meet New Rules, Says MacKay," Halifax *Chronicle Herald,* 18 April 2009.
40 European Aviation Safety Agency, *Type Certificate Data Sheet – EH-101,* 13 September 2006. Retrieved from http://www.easa.eu.int/ws_prod/c/doc/Design_Appro/Rotorcraft/TCDS%20EASA.R.013%20Agusta%20EH101-500%20and%20510.pdf.
41 Tiffany Crawford, "Helicopter Had Faulty Flotation Device: Crash Report," *Canwest News Service,* 18 June 2009.
42 James Bagnall, "Pressure Mounts on Key Supplier for DND Choppers," *Ottawa Citizen,* 5 March 2009.
43 David Pugliese, "Auditor General Won't Examine $5 Billion Maritime Helicopter Program," *Ottawa Citizen,* 23 January 2009, http://communities.canada.com/ottawacitizen/.
44 "More with Less," United Technologies Corporation Annual Report 2008, http://utc.com/.

Conclusion

Epigraphs: Douglas Bland, ed., *Canada's National Defence,* Vol. II: *Defence Organization* (Kingston: School of Policy Studies, 1998), 30; "Creating an Acquisition Model That Delivers," Vimy Paper 1, (Ottawa: Conference of Defence Associations Institute, April 2006), 20.

1 Gordon Davis, *The Maritime Helicopter Project: The Requirement for a Capable Multi-Purpose* Sea King *Replacement* (Halifax: Centre for Foreign Policy Studies, 2000), 12.

2 See Aaron Plamondon, "Casting Off the Yoke: The Transition of Canadian Defence Procurement within the North Atlantic Triangle, 1907-53" (MA thesis, Royal Military College of Canada, 2001).
3 James Fergusson, "In Search of a Strategy: The Evolution of Canadian Defence Industrial and Regional Benefits Policy," in *The Economics of Offsets: Defence Procurement and Countertrade,* ed. Stephen Martin (Amsterdam: Routledge, 1996), 107.
4 This and the following quotation were retrieved from Colonel W.N. Nelson, "The Need for a Viable Defence Industrial Base," *Canadian Defence Quarterly* 15 (1986).
5 Greig Stewart, *Shutting Down the National Dream: A.V. Roe and the Tragedy of the Avro* Arrow (Toronto: McGraw-Hill Ryerson, 1988), 2-3.
6 John Treddenick, "The Economic Significance of the Canadian Defence Industrial Base," Centre for Studies in Defence Resource Management, Royal Military College of Canada, Report 15, 1987, 1.
7 Dan Middlemiss, "Defence Procurement in Canada," in *Canada's International Security Policy,* ed. David B. Dewitt and David Leyton-Brown (Scarborough: Prentice Hall, 1995), 391.
8 John Treddenick, "The Economic Significance of the Canadian Defence Industrial Base," in *Canada's Defence Industrial Base: The Political Economy of Preparedness and Procurement,* ed. David G. Haglund (Kingston: R.P. Frye, 1988), 42.
9 J. Craig Stone, "Procurement in the 21st Century: Will the Present Process Meet the Requirement?" article submitted for the RMC War Studies Occasional Paper Series, 1999. See also James Fergusson, "In Search of a Strategy: The Evolution of Canadian Defence Industrial and Regional Benefits Policy," in Martin, *The Economics of Offsets,* 108.
10 Auditor General, *Report of the Auditor General of Canada to the House of Commons 1992* (Ottawa: Minister of Supply and Services Canada, 1992), 392.
11 David Underwood, "The Eyes and Ears of NSA," *Canadian Aviation* (1987): 21.
12 Colonel (retired) John Cody, CO of 423 Helicopter Squadron in 1987, e-mail to Aaron Plamondon, 5 March 2006.
13 This and the following quotation are from Colonel John Cody, e-mail to Aaron Plamondon, 13 January 2005.
14 Retrieved from an e mail from Major N.B. Barrett to N. Crawley and Colonel Drummond at the MHP PMO, 11 April 2001, AIA (AGU039876).
15 This and the following quotation were retrieved from Alan Williams, *Reinventing Canadian Defence Procurement: A View from the Inside* (Montreal and Kingston: McGill-Queen's University Press, 2006), 8, 2.
16 Ibid., 9-10.
17 Sharon Hobson, "Plain Talk: The Process of (Not) Acquiring Maritime Helicopters," *Canadian Naval Review* 4, 4 (2009): 39.
18 Colonel Brian Akitt, "The *Sea King* Replacement Project: A Lesson in Failed Civil-Military Relations," Canadian Forces College, 2003, 26; paper obtained from Colonel Akitt.
19 General (retired) George Macdonald, interview with Aaron Plamondon, 25 September 2005.
20 Jean Chrétien, *My Years as Prime Minister* (Toronto: Random House, 2007), 54.
21 See Aaron Plamondon, *Equipment Procurement in Canada and the Civil-Military Relationship: Past and Present* (Calgary: Centre for Military and Strategic Studies, 2008). For a great addition to the procurement debate by some of the only Canadian experts in the field see "Creating an Acquisition Model That Delivers," Vimy Paper 1 (Ottawa: Conference of Defence Associations Istitute, April 2006).

Selected Bibliography

Archival Sources

Library and Archives Canada (LAC)
Air Board Committee Minutes
Department of National Defence Records (RG24)
General (retired) Paul Manson Papers. (As of March 2006, these papers were not available to the public, but special access was provided to me by General Manson.)
Loring Christie Papers
Minutes of Imperial War Conference, 1917
Report of the Royal Commission on the Bren Machine Gun Contract

Directorate of History and Heritage, Department of National Defence
Minutes of Meetings of the Chiefs of Staff
Minutes of Meetings of the Naval Board
Minutes of Meetings of the Naval Policy Co-Ordinating Committee
Minutes of Meetings of the Naval Staff
Minutes of Meetings of the Policy Planning Co-Ordinating Committee
Minutes of Meetings of the Treasury Board
Minutes of Meetings of the Vice Chiefs of Staff Committee

Government Publications

Auditor General of Canada. "National Defence: In Service Equipment." Report to the House of Commons, December 2001.

–. *Report of the Auditor General of Canada to the House of Commons 1992.* Ottawa: Minister of Supply and Services Canada, 1992.

British Columbia. *Official Report of the Debates in the Legislative Assembly,* vol. 9, no. 5, 22 April 1993.

Canada. *Department of Defence Annual Report* for fiscal year ending March 1949. Ottawa: King's Printer, 1951.

–. Department of External Relations. *Documents on Canadian External Relations.* Vols. 1 and 2. Ottawa: Department of External Relations, 1967.

–. Department of National Defence. *Canadian Defence Budget.* Ottawa: Department of National Defence, 1963.

–. *Report of the Ad Hoc Committee on Defence Policy.* Ottawa: Department of National Defence, 30 September 1963.

–. *White Paper on Defence, 1947.* Ottawa: King's Printer, 1947.

–. *White Paper on Defence, 1949.* Ottawa: King's Printer, 1949.

–. *White Paper on Defence, 1987.* Ottawa: Queen's Printer, 1987.

–. *White Paper on Defence, 1994.* Ottawa: Queen's Printer, 1994.

–. House of Commons. *Debates*. Ottawa: Queen's Printer.

Canadian Liberal Party. *War Contract Scandals, as Investigated by the Public Accounts Committee of the House of Commons, 1915; also the Purchase of Boots, as Investigated by the Special "Boot Committee" Appointed by the House of Commons, Ottawa, 1915*. Ottawa: privately printed, 1915.

Directorate of Flight Safety. "Release for the Crash of *Sea King* Helicopter CH124-38 on February 2, 2006." http://www.airforce.forces.gc.ca/.

Federal Court of Canada. "Federal Court between AgustaWestland International Limited and Minister of Public Works and Government Services and Sikorsky International Operations, Inc." Received by request from the Federal Court of Canada, court file T-1605-04, filed 1 September 2004.

Kealy, J.D.F., and E.C. Russell. *A History of Canadian Naval Aviation, 1918-62*. Ottawa: Department of National Defence, 1965.

Pound, C.F.W. "The Defence Program and National Industrial Development." In *Defence Research Analysis Establishment Report* 34. Ottawa: Department of National Defence, 1973.

Stacey, C.P. *Arms, Men, and Government: The War Policies of Canada, 1939-45*. Ottawa: Queen's Printer, 1970.

Tucker, Gilbert. *The Naval Service of Canada, I: Origins and Early Years*. Ottawa: King's Printer, 1952.

Privately Obtained Documents

AgustaWestland International Limited. "Briefing Note: The Maritime Helicopter Project," 3 May 2001. Obtained by Colonel (retired) Laurence McWha.

Collenette, David. "Budget Impact: National Defence," February 1994. Acquired from former Assistant Deputy Minister (Materiel) Ray Sturgeon.

Hennessy, Michael. "The RCN and the Post War Naval Revolution, 1955-1964." Paper presented at the Canadian Historical Association Annual Conference, Charlottetown, 1992). Copy acquired from Michael Hennessy.

McDermott, John W., Lieutenant Colonel (retired). Speech written and given to Colonel John Orr. Acquired directly from Colonel Orr.

Multi-volume briefing package in regard to the NSA program, presented to Kim Campbell in 1993 after she became Prime Minister. Retrieved from the consulting firm TASC, Ottawa.

Stone, J. Craig. "Procurement in the 21st Century: Will the Present Process Meet the Requirement?" Article submitted for the Royal Military College (RMC) War Studies Occasional Paper Series, 1999. Acquired from Craig Stone.

Newspapers and Magazines

Air Force
Crowsnest
Halifax *Daily News*
Halifax *Chronicle-Herald*
Knight-Ridder Tribune
Maclean's
Montreal *Gazette*
The Financial Post
Le Devoir
Legion Magazine
The National Post
Ottawa Citizen
Ottawa *Sunday Sun*
Ottawa *Hill Times*
Regina *Leader-Post*
Sentinel
Summit
Toronto *Globe and Mail*
Toronto Star

Wednesday Report
Wings Magazine

Interviews
Colonel (retired) John Cody
Lieutenant General (retired) George Macdonald
General (retired) Paul Manson
Colonel (retired) Laurence McWha
Gordon Moyer
Colonel (retired) Lee Myrhaugen
Colonel (retired) John Orr
Ray Sturgeon

Web Pages
www.airliners.net
www.bloomberg.com
www.canoe.ca
http://collections.ic.gc.ca
www.defensedaily.com
www.dnd.ca
www.forces.gc.ca
http://jproc.ca
www.legis.gov.bc.ca
www.shearwateraviationmuseum.ns.ca
www.whl.co.uk

Secondary Sources
Adcock, Al. *H-3* Sea King *in Action*. Corrollton, TX: Signal Publications, 1995.
Archambault, Peter Michael. "The Informal Alliance: Anglo-Canadian Defence Relations, 1945-60." PhD diss., University of Calgary, 1997.
Armstrong, John G. "The Dundonald Affair." *Canadian Defence Quarterly* 11, 2 (Autumn 1981): 39-45.
Aronsen, Lawrence R. "Canada's Post War Re-Armament: Another Look at American Theories of the Military Industrial Complex." *Canadian Historical Association Historical Papers* 16, 1 (1981): 175-96.
–. "From World War to Limited War: Canadian-American Industrial Mobilization for Defence." *Revue internationale d'histoire militaire* 51 (1982): 208-45.
Axline, W. Andrew, ed. *Continental Community? Independence and Integration in North America*. Toronto: McClelland and Stewart, 1974.
Bercuson, David J. *Blood on the Hills: The Canadian Army in the Korean War*. Toronto: University of Toronto Press, 1999.
–. *True Patriot: The Life of Brooke Claxton, 1898-1960*. Toronto: University of Toronto Press, 1993.
Bercuson, David, and Barry Cooper. "Helicopter Replacement Fiasco." *Fraser Forum* (June 2003): 28-29.
Bercuson, David, Aaron Plamondon, and Ray Szeto. *An Opaque Window: An Overview of Some Commitments Made by the Government of Canada Regarding the Department of National Defence and the Canadian Forces*. Calgary: Canadian Defence and Foreign Affairs Institute, 2006.
Bland, Douglas, ed. *Canada's National Defence*, Vol. II: *Defence Organization*. Kingston: School of Policy Studies, 1998.
–, ed. *Transforming National Defence Administration*. Montreal and Kingston: McGill-Queen's University Press, 2005.
–. "A Unified Theory of Civil Military Relations." *Armed Forces and Society* 26, 1 (1999): 7-25.
Bothwell, Robert, and J.L. Granatstein. *Our Century: The Canadian Journey in the Twentieth Century*. Toronto: McArthur, 2000.

Boutilier, James A., ed. *The RCN in Retrospect, 1910-1968*. Vancouver: UBC Press, 1982.
Brassington, D.G. "The Canadian Development of VDS." *Maritime Warfare Bulletin*, commemorative ed. (1985).
Breemer, Jan. *Soviet Submarines: Design, Development, and Tactics*. Surrey, UK: Jane's Information Group, 1989.
Broad, Graham. "'Not Competent to Produce Tanks': The Ram and Tank Production in Canada, 1939-1945." *Canadian Military History* 11 (2002): 24-36.
Buckley, Brian. *Canada's Early Nuclear Policy: Fate, Chance, and Character*. Montreal and Kingston: McGill-Queen's University Press, 2000.
Byers, R.B. "Defence and Foreign Policy in the 1970s: The Trudeau Doctrine." *International Journal* 33 (1977-78).
Cafferky, Michael Shawn. "Towards the Balanced Fleet: A History of the Royal Canadian Naval Air Service, 1943-45." MA thesis, University of Victoria, 1989.
–. "Uncharted Waters: The Development of the Helicopter Carrying Destroyer in the Post-War Royal Canadian Navy, 1943-64." PhD diss., Carleton University, 1996.
Campagna, Palmiro. *Storms of Controversy: The Secret Avro* Arrow *Files Revealed*. Toronto: Stoddart, 1992.
Carnegie, David. *The History of Munitions Supply in Canada, 1914-1918*. London, UK: Longmans, Green, 1925.
Charlton, Peter. *Nobody Told Us It Couldn't Be Done: The Story of the VX10*. Ottawa: Peter Charlton, 1995.
Charlton, Peter, and Michael Whitby. *"Certified Serviceable"* Swordfish *to* Sea King*: The Technical Story of Canadian Naval Aviation by Those Who Made It So*. Gloucester, ON: CNATH, 1995.
Chartres, John. *Westland* Sea King. London: Ian Allan, 1984.
Cook, Glenn. *Vignettes of a Canadian Naval Aviator, 1955-1983: A Memoir*. Ottawa: self-published, 2005.
Cooling, Benjamin Franklin, ed. *War, Business, and World Military-Industrial Complexes*. New York: Kennikat Press, 1981.
Creighton, Donald. *Macdonald: The Young Politician – the Old Chieftain*. Toronto: University of Toronto Press, 1998.
Crosby, A. "Project Management in DND." *Canadian Defence Quarterly* 18, 6 Special no. 2 (1989): 59-63.
Cuff, R.D., and J.L. Granatstein. *Canadian-American Relations in Wartime: From the Great War to the Cold War*. Toronto: Hakkert, 1975.
Curleigh, Colin. "The New Maritime Helicopter: Reliability Will Be Crucial." *Canadian Defence Quarterly* 26, 4 (1997): 26-31.
Cuthbertson, Brian. *Canadian Military Independence in the Age of the Superpowers*. Toronto: Fitzhenry and Whiteside, 1977.
Davis, Gordon. *The Maritime Helicopter Project: The Requirement for a Capable Multi-Purpose* Sea King *Replacement*. Halifax: Centre for Foreign Policy Studies, 2000.
Davis, S. Mathwin. "It Has All Happened Before: The RCN, Nuclear Propulsion, and Submarines, 1958-68." *Canadian Defence Quarterly* 17 (1987): 34-40.
Dewitt, David B., and David Leyton-Brown. *Canada's International Security Policy*. Scarborough: Prentice-Hall, 1995.
Douglas, W.A.B., ed. *RCN in Transition, 1910-1985*. Vancouver: UBC Press, 1988.
Dow, James. *The* Arrow. Toronto: James Lorimer, 1997.
Duguid, A. Fortescue. *A Question of Confidence: The Ross Rifle in the Trenches*. Ottawa: Service Publications, 2000.
Dzuiban, Stanley. *Military Relations between the United States and Canada, 1939-1945*. United States Army in World War II Special Studies. Washington, DC: US Government Printing Office, 1959.
Edgar, Alistair D., and David G. Haglund. *The Canadian Defence Industry in the New Global Environment*. Montreal and Kingston: McGill-Queen's University Press, 1995.
English, John, and Norman Hillmer, eds. *Making a Difference: Canada's Foreign Policy in a Changing World*. Toronto: Lester Publishing, 1992.

Fetterly, Ross. "The Influence of the Environment on the 1964 Defence White Paper." *Canadian Military Journal* 5, 4 (2004-5): 47-54. http://www.journal.forces.gc.ca/.
Foster, J.A. *Muskets to Missiles: A History of Canada's Ground Forces*. Toronto: Methuen, 1987.
–. *Sea Wings: A History of Waterborne Defence Aircraft*. Toronto: Methuen, 1986.
Friedman, Norman. *The Post War Naval Revolution*. Annapolis: Naval Institute Press, 1986.
Gibbings, David. Sea King*: 21 Years Service with the Royal Navy, 1969-1990*. Seaborough Hill, UK: Skylark Press, 1990.
Gimblett, Richard. *Operation Friction: The Canadian Forces in the Persian Gulf, 1990-1991*. Toronto: Dundurn Press, 1997.
Godfrey, David. "Procuring Canada's New Helicopters: Still Firmly on the Rails, the Canadian Navy's New Shipborne Aircraft Program Has Survived Severe Cutbacks in Defence Spending." *Canada's Navy* (1991-92).
Gordon, Donald C. *The Dominion Partnership in Imperial Defence, 1870-1914*. Baltimore: Johns Hopkins University Press, 1965.
Granatstein, J.L. "The American Influence on the Canadian Military, 1939-1963." *Canadian Military History* 2, 1 (1993): 63-73.
–. *Arming the Nation: Canada's Industrial War Effort, 1939-45*. Ottawa: Council of Chief Executives, 2005.
–. *Canada's Army*: *Waging War and Keeping the Peace*. Toronto: University of Toronto Press, 2002.
–, ed. *Canadian Foreign Policy: Historical Readings*. Toronto: Copp Clark Pittman, 1986.
Hackman, Willem. *Seek and Strike: Sonar, Anti-Submarine Warfare, and the Royal Navy, 1914-54*. London: Science Museum, 1984.
Haglund, David G., ed. *Canada's Defence Industrial Base: The Political Economy of Preparedness and Procurement*. Kingston: R.P. Frye, 1988.
Hall, H. Duncan. *North American Supply*. London, UK: H.M. Stationery Office and Longmans Green, 1955.
Hall, H. Duncan, and C.C. Wrigley. *Studies of Overseas Supply*. London, UK: H.M. Stationery, 1956.
Harris, Stephen. *Canadian Brass: The Making of a Professional Army, 1860-1939*. Toronto: University of Toronto Press, 1988.
Hasek, John. *The Disarming of Canada*. Toronto: Key Porter, 1987.
Haycock, R.G., and B.D. Hunt. "Early Canadian Weapons Acquisition: 'That Damned Ross Rifle.'" *Canadian Defence Quarterly* 14, 3 (Winter 1984-85): 48-57.
–, eds. *Canada's Defence: Perspectives on Policy in the Twentieth Century*. Toronto: Copp Clark Pittman, 1993.
Haydon, Peter, and Don Macnamara. "Of Helicopters and Defence Policy." Canadian Institute of Strategic Studies Datalink 40, March 1993.
Hellyer, Paul. *Damn the Torpedoes: My Fight to Unify Canada's Armed Forces*. Toronto: Force 10 Publications, 1990.
Hillmer, Norman, and J.L. Granatstein. *Empire to Umpire: Canada and the World to the 1990s*. Toronto: Copp Clark Longman, 1994.
Hobson, Sharon. "Canada Amends Its Sikorsky *Cyclone* Contract." *Jane's Defence Weekly*, 30 December 2008.
–. "Plain Talk: The Process of (Not) Acquiring Maritime Helicopters." *Canadian Naval Review* 4, 4 (2009): 39-40.
–. "Winds of Change: *Cyclone* Brings Canada up to Date." *Jane's Navy International* (2005).
Horn, Bernd, ed. *Forging a Nation: Perspectives on the Canadian Military Experience*. St. Catharines, ON: Vanwell, 2002.
Huebert, Rob, Michael L. Hadley, and Fred W. Crickard, eds. *A Nation's Navy: In Quest of Canadian Naval Identity*. Montreal and Kingston: McGill-Queen's University Press, 1996.
Jockel, Joseph. *The Canadian Forces: Hard Choices, Soft Power*. Toronto: Canadian Institute of Strategic Studies, 1999.
–. *No Boundaries Upstairs: Canada, the United States, and the Origins of North American Air Defence, 1945-1958*. Vancouver: UBC Press, 1987.

Johnston, William. "Canadian Defence Industrial Policy and Practice: A History." *Canadian Defence Quarterly* 18, 6 Special no. 2 (June 1989): 21-28.
Jordan, John. *Soviet Submarines, 1945 to Present.* London: Arms and Armour Press, 1989.
Law, Clive M. *Making Tracks: Tank Production in Canada.* Ottawa: Service Publications, 2001.
Low, Floyd. "Canadian Militia Policy, 1885-1914." *Canadian Defence Quarterly* 11, 2 (Autumn 1981): 29-38.
Lukasiewicz, Julius. "Canada's Encounter with High-Speed Aeronautics." *Technology and Culture* 27, 2 (1986): 223-61.
Lund, W.G.D. "The Rise and Fall of the Royal Canadian Navy, 1945-64: A Critical Study of the Senior Leadership, Policy, and Manpower Management." PhD diss., University of Victoria, 1999.
Lynch, Thomas. "Canada's NSA Program: And Then There Was One." *Wings, Canada's Navy Annual* (1988-89).
–. "Naval Shipborne Aircraft: Rotary Flight after the *Sea King.*" *Canada's Navy* (1986).
–. "New Shipborne Helicopter Program." *Wings, Canada's Navy Annual* (1987-88).
–. "Stuffing NSA: DND and Canadian Industry Gear Up to Provide Comprehensive Mission Suite." *Wings, Canada's Navy Annual* (1987-88).
MacAleese, J.W. "The CH-146 *Griffon*: Underrated and Over Criticized." Paper written at the Canadian Forces College, 2001. http://wps.cfc.forces.gc.ca/.
Manson, Paul. "The CF-18 *Hornet*: Canada's New Fighter Aircraft." *Canadian Defence Quarterly* 10 (1980).
–. "Procurement Cycle Growth: The Race between Obsolescence and Acquisition of Military Equipment in Canada, 1960 to the Present." Paper presented at the Canadian Institute of Strategic Studies Seminar, July 2005. http://www.cda-cdai.ca/.
Marchant, J.P., and Chris Bullock, eds. *Perspectives on War: New Views on Historical and Contemporary Security Issues.* Calgary: Centre for Military and Strategic Studies, 2003.
Martin, Stephen, ed. *The Economics of Offsets: Defence Procurement and Countertrade.* Amsterdam: Routledge, 1996.
McAndrew, William. "The Early Days of Aircraft Acquisition in Canadian Military Aviation." *Canadian Defence Quarterly* 12 (1982).
McKay, R.A, ed. *Canadian Foreign Policy 1945-54: Selected Speeches and Documents.* Toronto: McClelland and Stewart, 1970.
McLin, Jon B. *Canada's Changing Defence Policy, 1957-63: The Problem of a Middle Power in Alliance.* Baltimore: Johns Hopkins University Press, 1967.
Middlemiss, Danford. "Canadian Defence Funding: Heading towards Crisis?" *Canadian Defence Quarterly* 21 (1991): 13-20.
–. "A Pattern of Cooperation: The Case of the Canadian-American Defence Production and Development Sharing Arrangements, 1958-1963." PhD diss., University of Toronto, 1975.
Milner, Marc. *Canada's Navy: The First Century.* Toronto: University of Toronto Press, 1999.
Morton, Desmond. *Canadian Military History: From Champlain to the Gulf War.* Toronto: McClelland and Stewart, 1992.
–. *Ministers and Generals: Politics in the Canadian Militia, 1868-1904.* Toronto: University of Toronto Press, 1970.
–. *Understanding Canadian Defence.* Toronto: Penguin Canada, 2003.
–. *When Your Number's Up: The Canadian Soldier in the First World War.* Toronto: Random House, 1994.
Morton, Desmond, and J.L. Granatstein. *Marching to Armageddon: Canada and the Great War 1914-1919.* Toronto: Lester and Orpen Dennys, 1989.
Murray, Robert. *Sikorsky HO4S-3 (S-55)* Horse, *Royal Canadian Navy (RCN).* Ottawa: Canadian Aviation Museum, n.d. http://www.aviation.technomuses.ca/.
Nelson, W.N. "The Need for a Viable Defence Industrial Base." *Canadian Defence Quarterly* 15 (1986).
Newman, Peter C. *The Secret Mulroney Tapes: Unguarded Confessions of a Prime Minister.* Toronto: Random House Canada, 2005.

Officer, Lawrence H., and Lawrence B. Smith. "Stabilization in the Postwar Period." In *Canadian Economic Problems and Policies*, ed. Lawrence H. Officer and Lawrence B. Smith. Toronto: McGraw-Hill, 1970.

Orr, John L., ed. *"With Eagle Wings," 423: A Canadian Squadron in Peace and War*. 423 Squadron Sixtieth Anniversary History Project. Halifax: Print Atlantic, 2002.

Pettipas, Leo. *Canadian Naval Aviation, 1945-68*. Winnipeg: Pettipas, 1990.

Phillips, Roger F. *The Ross Rifle Story*. Sydney, NS: John A. Chadwick, 1984.

Pickersgill, J.W., and D.F. Forster. *The Mackenzie King Record*. Vol. 4. Toronto: University of Toronto Press, 1960.

Pierce, S.D., and A.F.W. Plumptre. "Canada's Relations with War Time Agencies in Washington." *Canadian Journal of Economics and Political Science* 11, 3 (1945): 402-19.

Plamondon, Aaron. "Casting Off the Yoke: The Transition of Canadian Defence Procurement within the North Atlantic Triangle, 1907-53." MA thesis, Royal Military College of Canada, 2001.

–. *Equipment Procurement in Canada and the Civil-Military Relationship: Past and Present*. Calgary: Centre for Military and Strategic Studies, 2008.

Porter, Gerald. *In Retreat: The Canadian Forces in the Trudeau Years*. Ottawa: Deneau and Greenberg, 1978.

Preston, Richard A. *Canada and "Imperial Defence": A Study of the Origins of the British Commonwealth's Defense Organization, 1867-1919*. Durham: Duke University Press, 1967.

–. *The Defence of the Undefended Border: Planning for War in North America 1867-1939*. Montreal and Kingston: McGill-Queen's University Press, 1977.

Priestley, S.T. "Politics, Procurement Practices, and Procrastination: The Quarter-Century *Sea King* Helicopter Replacement Saga." *Canadian-American Strategic Review*. http://www.casr.ca/ft-mhp1.htm.

Rawling, Bill. *Surviving Trench Warfare: Technology and the Canadian Corps, 1914-1918*. Toronto: University of Toronto Press, 1992.

Richter, Andrew. *Avoiding Armageddon: Canadian Military Strategy and Nuclear Weapons, 1950-63*. Vancouver: UBC Press, 2002.

Roy, Reginald H. *For Most Conspicuous Bravery: A Biography of Major-General George R. Pearkes, V.C., through Two World Wars*. Vancouver: UBC Press, 1977.

Sarty, Roger. *The Maritime Defence of Canada*. Toronto: Canadian Institute of Strategic Studies, 1996.

Shadwick, Martin. "Replacing the *Sea King:* Canada Examines the Need to Replace Its *Sea Kings* with a New ASW Helicopter." *Wings, Canada's Navy Annual* (1985).

Simpson, Erika. *NATO and the Bomb: Canadian Defenders Confront Critics*. Montreal and Kingston: McGill-Queen's University Press, 2001.

Snowie, J. Allan. *The Bonnie: HMCS* Bonaventure. Erin, ON: Boston Mills Press, 1987.

Sokolsky, Joel J., and Joseph T. Jockel. *Fifty Years of Canada-United States Defence Cooperation: The Road from Ogdensburg*. Lewiston: Edwin Mellen Press, 1992.

Soward, Stuart E. *Hands to Flying Stations: A Recollective History of Canadian Naval Aviation*. Vol. 2. Victoria: Neptune Developments, 1995.

Stacey, C.P. *Arming the Nation: Canada's Industrial War Effort, 1939-45*. Ottawa: Council of Chief Executives, 2005.

–. *The Military Problems of Canada: A Survey of Defence Policies and Strategic Conditions Past and Present*. Toronto: Canadian Institute of International Affairs/Ryerson Press, 1940.

Stanley, George. *Canada's Soldiers: The Military History of an Unmilitary People*. Toronto: Macmillan, 1974.

Stewart, Greig. *Shutting Down the National Dream: A.V. Roe and the Tragedy of the Avro* Arrow. Toronto: McGraw-Hill Ryerson, 1988.

Treddenick, John. "The Economic Significance of the Canadian Defence Industrial Base." Centre for Studies in Defence Resource Management Report 15. Kingston: Royal Military College of Canada, 1987.

Underwood, David. "The Eyes and Ears of NSA." *Canadian Aviation* 60 (1987).

Wakelam, Randall. "Flights of Fancy: RCAF Fighter Procurement 1945-1954." MA thesis, Royal Military College of Canada, 1997.

Yost, William J. *Industrial Mobilization in Canada*. Ottawa: Conference of Defence Associations, 1983.
Zimmerman, David. *The Great Naval Battle of Ottawa: How Admirals, Scientists, and Politicians Impeded the Development of High Technology in Canada's Wartime Navy*. Toronto: University of Toronto Press, 1989.

Index

Note: "(i)" after a page number indicates an illustration.

Printed and bound in Canada by Friesens
Set in Stone by Artegraphica Design Co. Ltd.
Copy editor: Dallas Harrison
Proofreader: Dianne Tiefensee
Indexer: Natalie Boon

ENVIRONMENTAL BENEFITS STATEMENT

UBC Press saved the following resources by printing the pages of this book on chlorine free paper made with 100% post-consumer waste.

TREES	WATER	SOLID WASTE	GREENHOUSE GASES
7 FULLY GROWN	3,199 GALLONS	194 POUNDS	664 POUNDS

Calculations based on research by Environmental Defense and the Paper Task Force.
Manufactured at Friesens Corporation